DNA and Destiny

*Nature and Nurture
in Human Behavior*

DNA and Destiny
Nature and Nurture in Human Behavior

R. GRANT STEEN

PLENUM PRESS • NEW YORK AND LONDON

Library of Congress Cataloging-in-Publication Data

On file

ISBN 0-306-45260-X

© 1996 R. Grant Steen
Plenum Press is a Division of Plenum Publishing Corporation
233 Spring Street, New York, N.Y. 10013-1578

10 9 8 7 6 5 4 3 2 1

Printed in the United States of America

*Be patient toward all that is unresolved in your heart
and try to love the questions themselves.*—Rainer Maria Rilke

Acknowledgments

My wife, Wil O'Loughlin, has been a wonderful partner during the draining process of writing, and she and my children, Alena and Mariel, have provided love, light, warmth, and laughter, as well as a healthy dose of perspective. My parents, Noreen and Ralph Steen, have been supportive and enthusiastic throughout the process of clarifying confusion, identifying ignorance, and delimiting real knowledge. Colleagues at the University of California, Los Angeles, have played a large role in teaching me to think like a scientist while writing like a person: Dr. Leonard Muscatine still serves as a model of clarity in writing; Dr. George Laties is an example of the eternal youth of the questioning mind; and Dr. Margaret McFall-Ngai remains one of the best scientists I know. I also want to thank Matthew Galvez and Dr. Linda Porter, who encouraged me to follow this path; Dr. Barry Fletcher and Dr. June Taylor, who make it possible to continue; and Linda Regan and Jeanne Fredericks, who have helped it all to happen.

Contents

1

Genes and Human Behavior

Animal breeders, who bred and selected for loyalty in dogs, or speed in horses, or docility in cattle, were probably the first to notice the powerful influence of heredity, even if they didn't know the word or understand the principles. Certainly all successful breeders realized that much could be known about the traits of a newborn animal by having a thorough familiarity with the bloodlines of that animal. The bloodlines of racing horses in England have been recorded in the *Stud Book* for more than 200 years. The impetus for keeping this record for so many years is the sure knowledge that a cross between two slow horses will virtually never produce a fast horse. This simple perception is an affirmation that, to at least some extent, DNA is destiny. The first *Stud Book*, published in 1793, listed almost 200 horses available for sire, but only 3 of those horses have had offspring through all of the intervening years. Thus, every thoroughbred racing horse in the world today can trace its ancestry back to one of these 3 horses. Yet, even though the bloodline of each racing horse in the world is known, in some cases back for more than 20 generations, breeding a faster racehorse remains somewhat a matter of chance. While DNA may decree destiny, the destiny it decrees is apparently not a fixed and immutable one.

The medieval preoccupation with the bloodlines of nobility, or the Victorian emphasis on noble breeding as a prerequisite for inclusion in polite society, both may originate from the perception that genes can determine the future as well as the past.

Similarly, the current fascination with genealogy may arise from the realization that an above-average ancestor can be a source of good genes, as well as a source of good cocktail-party conversation. And when a modern father asks his daughter if her new boyfriend "comes from a good family" he is really looking for a rough indication of what the future may bring. Perhaps if the new boyfriend comes from a long line of intelligent, well-balanced, and successful forebears, he will not turn out to be the axe-murderer he resembles.

The first recorded experiment designed to test the question of whether nature or nurture is more important in human development occurred in the 13th century. King Frederick II of Germany became curious about what kind of language would develop among children reared without any exposure to preexisting language.[1] He was curious as to whether children would teach themselves the Hebrew language, which was the oldest language extant in Europe at the time, or one of the more recent languages such as Greek or Latin. Therefore, he ordered foster mothers to rear and care for a series of children without speaking to them, or exposing them to language in any way. But this experiment was a terrible failure, since all of the children died fairly quickly. Apparently they could not live without the bonding that is fostered through language.

Six centuries later in England, Francis Galton studied the heritability of genius. Galton began by identifying the most celebrated men of his day, men who were famous scientists, authors, judges, musicians, military leaders, clerics, and statesmen. He then searched the family trees of these men to determine whether or not they were related to other eminent men of the past or present. Galton set the most famous person in each family as the referent, and determined the likelihood that one famous person would be related to another by blood. He found that a famous man was far more likely to have a famous son or brother than to have a famous cousin. This implies that whatever qualities made the referent individual famous are likely to be present in near relatives as well. Galton's study, published in 1869 as a book entitled *Hereditary Genius: An Inquiry into its Laws and Consequences,* became very widely read, and it inspired generations of scientists to believe that heredity has much to say about the course of the future.

In 1871, Charles Darwin wrote in *The Descent of Man* that ". . . in regard to mental qualities, their transmission is manifest in our dogs, horses, and other domestic animals. Besides special tastes and habits, general intelligence, courage, bad and good tempers, etc., are certainly transmitted." This was a key observation for Darwin, because it implied that mental and physical qualities both could be selected for by a breeder interested in changing the traits of an animal. It was only one step further from this idea to the idea that nature itself could act as an agent of selection. In fact, Darwin used his observations on the breeding of domesticated animals as one of the central arguments in *The Descent of Man,* since this was something he and every other Englishman of his day was familiar with. Natural selection and the heritability of traits remain the central tenets of the theory of evolution, since new traits must be inherited for evolution to occur. Darwin's emphasis on the importance of heredity in determining both physical and mental traits persuaded generations of scientists to see the invisible hand of genetics at work everywhere, and this bias continues to the present day.

The flush of recent successes in molecular biology has focused public attention on that most complex and mysterious of all chemicals, DNA. Every feature of our physical being is coded for by DNA, and many scientists would argue that virtually every aspect of humanity is similarly encoded. The Human Genome Project, which proposes to sequence the entire human complement of DNA, is an acknowledgment that to know the code is, at least in part, to know the bearer of the code. The hype and hard sell surrounding this project has by now convinced many people that our future is indelibly written in our genes. Scientists have been reluctant to make explicit how large the gap is between knowing the sequence of the human genome and understanding the various ramifications of that sequence. Consequently, there is now a perception that we are closing in on a molecular understanding of the human condition. And most people are aware that, though genes can determine good health and above-average intelligence, genes are also responsible for many undesirable traits. There is a growing concern, to the point of consensus, that genes can predict a limited future for certain unfortunate individuals. This fatalism is a recurrent theme of the recent book *The Bell Curve*, as it makes the argument that class structure in the United States is largely determined, even defined by intelligence. However, the question of what determines intelligence is never adequately dealt with: if nature determines intelligence, then the defeatist theme of the book is perhaps appropriate; but if nurture determines intelligence, then giving up on the disadvantaged will eventually be equivalent to giving up on our society.

In a sense, this is the dark side of the revolution in our understanding of genetics; a fatalism born of the conviction that one cannot rise above one's genetic station. But this genetic fatalism is an overreaction; molecular biology and molecular genetics may be king, but there are troubles in the kingdom. In fact, molecular biology, evolutionary theory, and chaos theory appear to be somewhat at odds with one another. Molecular biology is concerned, at least in part, with the molecular machinery of heredity. Two basic tenets of molecular biology are that the function of the whole can be understood completely in terms of the parts, and that the machinery of heredity is very nearly infallible. Molecular biologists thus adhere to a somewhat static world view, in that change is generally not permitted to enter the picture. On the other hand, evolutionary theory is concerned with change and the consequences of inheritance. A basic tenet of evolutionary theory is that the evolution of organisms is inevitable, because the environment is constantly changing and because the machinery of heredity is imperfect. Finally, chaos theory is concerned with discerning the underlying structure of seemingly random objects or events. A basic tenet of chaos theory is that it is essentially impossible to define the whole by studying the parts, because we have an inadequate grasp of the parts and because conditions are changing far too rapidly. Thus, there seem to be some basic philosophical differences among three of the most dominant and important theories in the life sciences today. Apparently, molecular biology is king, and the king is fully clothed, but there are some rather large rends in those garments.

For our purposes, the tension between these several disciplines can be summarized in a simple question: If genes preordain the future, why then does the future remain so unpredictable? To put it another way, if we are all extraordinarily precise machines, functioning in a way completely determined by our genes, why is continual change in the human condition seemingly so inevitable? It may seem that nature has overwhelmed nurture, and that upbringing essentially has no importance. But to argue that genes are destiny is surely as naive as to argue that the environment alone determines the growth and development of a child. To understand the human condition better we must develop a keener appreciation for the subtle interaction between nature and nurture. As the late psychologist Donald O. Hebb of McGill University put it, "Heredity determines the range through which environment can modify the individual." In part, this simply concedes that you can't make a silk purse out of a sow's ear, but there is more to it than that. This simple quote also acknowledges that there is an ongoing interplay between the genes and the environment, such that we are each a product of both forces. We are an amalgam of ability and opportunity, a result of chance, choice, and necessity. There is a continual tension between the possible and the actual, with the possible determined by the genes, and the actual determined by the environment.

How Genes Influence Simple Traits

Deoxyribonucleic acid (DNA) is the blueprint of being; an enormously large and complex molecule that encodes all of the genetic information necessary to build a life. Each cell of the human body is thought to contain a complete copy of the genetic information needed to construct, not just the cell, but the entire body. DNA can be thought of, and in fact is often written, as a string of letters. The sequence of letters is equivalent to a code that determines the structure of the body's proteins. For each protein there is one gene; for each gene, one protein. What makes each of us a kind of biological singularity is that we are a unique collection of proteins, coded for by a unique collection of genes carried on the DNA molecule. Of course, our genes are a blend of what is inherited from each parent, so we share 50% of our biological identity with each parent. But the way in which our particular set of genes is regulated and expressed makes us completely unique.

For the sake of safety and stability, individual pieces of DNA are usually wrapped in protein and coiled into a structure known as a chromosome. Each cell of the human body has 23 pairs of chromosomes which together can specify the structure and function of the vast number of cells in our body. Each chromosome pair is a matched set, in the sense that each of the chromosomes in the pair contains the genetic information necessary to make the same set of proteins. However, two chromosomes in a pair are never identical to each other; the precise code for how to synthesize a particular protein can vary substantially between chromosomes in the pair. This is a tribute to the genetic differences that can be harmonized in physical union, since half of the chromosomes are inherited from the mother and half from the father.

In a sense, each cell actually contains twice as much information as is necessary for life. This redundancy may serve many purposes, but it certainly leads to greater genetic diversity among humans. It may also serve as a kind of genetic insurance; if an abnormal gene is inherited from one parent, there is a good chance that it can be compensated for with a normal gene inherited from the other parent. Alternatively, since one gene codes for one protein, two genes for the same protein can code for two slightly different versions of the same protein. While these two versions of the same protein would both serve the same function, they might do so in a slightly different fashion.

In the simplest possible case, one gene controls the expression of one protein, and that one protein controls the expression of a particular trait. This is referred to as "single trait" inheritance, and is by far the simplest case dealt with in genetics. Needless to say, single trait inheritance is not all that common; a great deal of evidence shows that the vast majority of human traits are controlled by several to many different genes. Yet single trait inheritance is important nonetheless, because some human traits do follow this simple pattern.

As an example of single trait inheritance, let us look briefly at sickle-cell disease, a devastating illness that reduces the ability of blood to deliver oxygen to the tissues. Sickle-cell disease most commonly afflicts those of Central African origin, but it can also affect peoples of Arabic or Indo-European origin. The disease is simple enough that its genetic basis has been understood for more than 40 years; a mutation affecting a single blood protein is the problem. This mutation reduces the ability of hemoglobin in the bloodstream to bind to oxygen, and causes red blood cells that contain the mutant hemoglobin to form abnormal sickle-shaped cells whenever oxygen is in short supply. The sickled red blood cells clog up the capillaries, so that blood flow to tissues can be temporarily interrupted. As simple as it sounds, the disease can have a devastating effect on even the very young. The consistent shortage of oxygen resulting from sickle-cell disease can cause massive and fatal strokes in children as young as 1 or 2 years old, and it commonly causes impaired growth and development, increased susceptibility to infection, congestive heart failure, debilitating and frequent pain crises, and chronic damage to virtually every organ of the body.

Yet sickle-cell disease is typically seen only in patients who have inherited two mutant copies of the gene for hemoglobin. A person who is fortunate enough to inherit one normal copy of the gene will not suffer the symptoms of sickle-cell disease, although they can pass the disease on to their children. If a person has one normal and one mutant copy of the hemoglobin gene, half of the hemoglobin in their bloodstream will be mutant; yet under most circumstances the normal hemoglobin can completely compensate for the abnormal hemoglobin and there are no symptoms at all. In general, any individual with two identical copies of a particular gene is referred to as being homozygous for that gene, whether or not the gene is normal. Someone with two different copies of a particular gene is referred to as being heterozygous for that gene. Thus, a person who carries the sickle-cell trait but does not suffer the disease is said to be heterozygous for sickle-cell disease.

How Genes Influence Complex Traits

A great deal of evidence shows that the vast majority of human traits are controlled by many different genes. Basically, any trait that varies over a broad range of expression is believed to be controlled by several to many different genes, in a process called multigenic inheritance. Classic examples of traits that show a continuous range of variation caused by multigenic inheritance include height, weight, and intelligence.

Height could not possibly be controlled by a single gene because, if it were, there would be only three different heights possible. If only one gene was involved, everyone homozygous for the tall gene would be tall, everyone homozygously lacking the tall gene would be short, and all those heterozygous for the tall gene would be the same intermediate height. But humans vary in height over an extraordinarily broad range. According to *The Guinness Book of Records,* the tallest man ever recorded was over 8 feet 11 inches tall and the shortest man, under 1 foot 11 inches. The fact that even two people fall so far out of the normal range of height suggests that a great many genes may be involved in determining human height.

But how do different genes interact with one another to determine human height? Although the answer to this question is not yet known, we can speculate that adult height must be the result of many separate genetic interactions. Without a doubt, someone who is tall must either grow quickly or must keep growing for a long time; both of these traits are probably under genetic control. Similarly, above-average height is more than likely associated with above-average synthesis of growth hormone or above-average sensitivity to growth hormone, both of which must be hereditary. Yet growth in height is possible only if a person has a good supply of food and the ability to digest that food efficiently; while the former is not under genetic control, the latter is certainly heritable. The ability to digest food efficiently would be useless, however, if there was not also a system in place to efficiently move nutrients to where they are needed; it is possible that tall persons have inherited some physiological advantage that makes them better able to use nutrients for growth. Similarly, a person can only be tall if the mass of their bones is sufficient to support the additional weight, and if the muscles are strong enough to move those bones. This may seem like a trivial problem but, from a biological standpoint, coordinating the growth of different tissues is actually not all that simple. Just the traits listed above would probably involve at least six different genes, and there are almost certainly many, many other traits involved in achieving tall stature. The problem is that it is, as yet, virtually impossible to determine how many genes are involved, let alone what the genes actually do.

The term *multigenic inheritance* is thus quite vague, indicating that no one knows exactly how many genes are involved in determining a human trait that shows a wide range of variation. As we saw from the example of sickle-cell disease, the effects caused by a single gene can be extremely complex and difficult to decipher. The situation is infinitely more complex and more difficult to decipher if a trait is the product of many different genes acting together. Each gene can have a subtle effect itself, different genes can have additive and interactive effects among themselves, and some genes may even cancel out the effect of other genes. And, of course, the environment has a powerful effect on the way different genes are expressed.

Behavior as a Complex Multigenic Trait

Behavior is surely one of the most complex and subtle of all human traits. Psychologists and psychiatrists often have trouble even defining and classifying behavior, so it goes without saying that geneticists have had a hard time determining the genetic basis of behavior.[2] If behavior is the sum and substance of our responses, both internal and external, to the stimuli presented to us by the environment, it may not be immediately obvious that behavior is inherited at all. But there is very strong evidence that most animal behavior is heritable, and there is growing evidence that human behavior is also heritable. As a rule, the more primitive the animal, the more likely it is that all or most of the behavior is inflexibly determined by genes.

In the fruit fly *Drosophila*, behavioral mutants are known to exist. This humble creature is one of the best-understood organisms on earth, having been a favorite experimental subject with geneticists for many years; the flies are easy to breed, they reproduce very quickly, and they have been closely studied for more than half a century. Certain flies are known to have a mutation that has an effect on some aspect of their behavior. For example, the very stereotypical sequence of events leading to copulation in fruit flies occurs only when both the male and female fly are behaviorally normal.[3] Several fruit fly mutations have been identified, each of which has an impact on the courtship pattern or on the eventual likelihood of success in reproduction. Some mutations affect the male and some affect the female, but all are somehow disruptive to the normal pattern of courtship. For example, male *white* eye mutants are known to be infrequently successful in their courtship, perhaps because this mutation affects the fly's ability to see. A male *white* eye mutant is often unable to follow a retreating female fly as well as can a normal red-eyed fly, and his initial overtures are seldom sufficient to induce a female to copulate immediately. Conversely, the *spinster* mutation affects the female, causing her to be unreceptive to all male sexual advances, and to respond to those advances with an exaggerated form of behavior indicating rejection.

Male fruit flies in search of a mate often sing a courtship song, in much the same way as lonesome crickets do. This courtship song is actually not vocalized, but rather is produced as the male fly vibrates his wings in a characteristic pattern that cannot be heard by humans. The courtship song is unique for each different type or species of fruit fly, and it has a powerful aphrodisiacal effect on female fruit flies of the same species. The particular pattern of song behavior in a *Drosophila* male is actually genetically determined, and mutations can be made that render a male unable to sing a proper courtship song and hence unable to attract a female fly of the same species.[4] Strangely, the courtship song is determined by only a few genes and song behavior can actually be transferred between different species by transferring these genes. This was proven a few years ago by scientists working with two species of *Drosophila* whose song patterns are ordinarily quite distinct from one another. A small piece of DNA, containing a gene called *period,* was isolated from one species; this gene is believed to control the rhythm or periodicity of the courtship song. This small piece of DNA was then transfected from one to the other species of fruit fly, and males bearing the altered gene were allowed to mature to adulthood. When these artificially mutated flies were tested, they were found to sing a courtship song very similar to that of the flies from which the DNA was taken. Evidently, fruit fly courtship behavior is not only carried on the genes, but the behavior can be transferred when the genes are inserted into another fly.

The fact that fruit fly mating behavior can be passed between flies like a virus is interesting, but it doesn't speak for human behavior. Although human behavioral mutants seem to be evident almost daily in the newspaper, scientific evidence is only now beginning to accumulate suggesting that virtually all human behavior has some measure of genetic contribution. But human behavioral patterns can be so strongly affected by external events that it is very difficult to "prove" that some behavioral pattern has a genetic component. Humans are very much harder to study than fruit flies, because they are so complex and so subtle, because there are an almost limitless range of behavioral options open to humans, and because it is impossible to do the controlled breeding experiments necessary to really understand the inheritance of behavior. Yet perhaps by studying simple organisms such as the fruit fly, scientists can gain some understanding of the programmed or genetically determined part of human behavior.

It is very difficult to imagine how genes, which simply and only carry the instructions to make proteins, could possibly influence a person to be intelligent, let alone homosexual, extroverted, or alcoholic. Yet the fact that we cannot understand how this happens does not mean that it cannot happen. We must expect that if inheritance of human behavior is possible, then the pattern of inheritance will be multigenic, because human behavior seems far too complex and too variable to be inherited otherwise. While it is hard to imagine that aesthetic appreciation could be heritable, to believe that there is a single gene for art appreciation would exceed the bounds of credibility. And any trait that is passed on by multigenic inheritance will necessarily be difficult to understand.

Genes, Traits, and Evolution

Before we continue with a discussion of the heritability of human behavior, something must be said about the single most important, most unifying principle in biology, namely, the concept of evolution. Evolution has come under some fire lately, from religious groups and the popular press, but the truth of the matter is that it remains the central concept in biology. None of the diversity we see around us can be explained in any other way, and there is a great deal of evidence that evolution is a strong creative force in the world around us.

To clear away the intellectual clutter that surrounds it, evolution can be reduced to a set of very simple tenets, each of which is consistent with day-to-day experience. If all of these tenets are true, then evolution is essentially inevitable. These basic tenets are as follows:

1. Variation exists within a species. Certainly human appearance spans a broad range of variation, and one can only assume that other traits that are less evident also span an equally broad range of variation. The range of variation possible within a species can be seen even more clearly in dogs: dachshunds and Great Danes are the same species and, barring certain obvious mechanical difficulties, these animals are capable of breeding to produce viable offspring.

2. Certain traits provide a competitive advantage. Clearly, a cat with claws has an advantage over a cat with no claws; in the same sense, large or strong claws provide a competitive advantage over small or weak claws. In human society, great intelligence or great physical stamina provide one with a great competitive advantage over someone else not so blessed.

3. Certain traits can be passed down from one generation to the next. A cross between two large dogs tends to produce a pup that will grow into a large adult. A mating between two cats with strong claws will likely produce a kitten with strong claws. And, as Galton showed, eminent men tend to have eminent children, perhaps because they have passed down some trait that enables eminence.

4. If a certain trait provides a competitive advantage, the frequency of that trait will likely increase over time.

Thus, cats with weak claws may be less able to hunt and less able to survive to adulthood; if they never have a chance to breed, then there will necessarily be fewer cats with weak claws in the next generation. Alternatively, a cat with strong claws may be able to mate more frequently, or it may be better able to feed kittens and raise them to adulthood. The total number of cats with strong claws may thus increase over time, so that they come to represent a greater proportion of all cats; this change in frequency of a trait over time is the essence of natural selection.

This is evolution in a nutshell. As traits change in frequency, so do the genes that specify those traits. Thus, there may, over time, be an increase in the frequency of genes for strong claws among cats, or of genes for intelligence among humans. As long as traits form the basis for natural selection, then a change in the frequency of traits is inevitable. This is because, if some trait gives an organism a competitive advantage, then that trait will be passed on and will eventually become more and more commonplace. In nature, death is the final arbiter of whether a trait is successful or not. Unless the strong die as easily as the weak, or the intelligent are no more successful than the dull, then evolution must occur. Evolution is simply inevitable unless all creatures die in a completely random fashion.

Behavioral genetics tacitly assumes that evolution occurs, and that certain behavioral traits can form a basis for natural selection. Certainly it is easy to see how cooperative behavior in honeybees could be selected for; if bees in a hive fail to cooperate, they will certainly die. It is perhaps less obvious that human behavioral traits could also form a basis for selection. We may need to look back to our distant ancestors to see the circumstances under which evolution of behavioral traits occurred. A group of primitive human ancestors unable to form a cooperative hunting band would probably have been doomed to extinction. The ability to form such a hunting band must have required a broad range of social skills that would then have become selectable traits. Natural selection for the ability to form successful hunting bands may therefore account for the evolution of social interaction, language, altruism, tool use, and a range of other traits often seen as human.

How Does the Environment Affect the Expression of Genetic Traits?

The disease phenylketonuria (PKU) provides one of the clearest examples of how nature and nurture can interact to control the expression of a human trait. PKU may be a familiar term to many because all children in the United States are routinely screened for this condition at birth, even though it afflicts fewer than one child in 20,000. The disease is caused by an inherited defect of one enzyme that helps to metabolize phenylalanine, an amino acid that is a common element in the diet. For the disease to develop, a patient has to be homozygous for the mutation, since a normal copy of the gene is able to compensate for a mutated copy of the gene. The inability to properly metabolize phenylalanine causes this chemical to accumulate in the bloodstream, where it eventually reaches toxic concentrations. Without treatment, the eventual result of this accumulation is mental retardation, which develops slowly over the course of several years. Children with untreated PKU fail to mature properly and eventually must be institutionalized because of profound retardation, seizures, and various physical and psychiatric problems. The fact that all children are screened for PKU, even though it is very rare, is a measure of how devastating the illness can be.

Although PKU is completely heritable, the trait is not expressed if phenylalanine is eliminated from the diet. This is easier said than done, though; phenylalanine is a building block of protein, so it is a required part of the diet. But children with PKU are put on a diet to reduce their intake of phenylalanine to a level that permits them to grow properly without the symptoms that result from phenylalanine accumulation. This diet replaces normal dietary protein with an artificial amino acid mixture low in phenylalanine; this mixture is supplemented with a small amount of natural food to meet the dietary requirement for phenylalanine. Although this condition is quite rare, it clearly demonstrates that human traits can result from complex and subtle interactions between genes and the environment.

A trait like PKU would normally be expected to disappear over time because natural selection is so harsh. In fact, until very recently, virtually all children with PKU died before they could have children. PKU may actually already be in the process of disappearing, since it is so rare. But, by this logic, there should be few or no harmful mutations present in humans, because natural selection acts against all of them. So, why are there many extremely harmful mutations that are more common than PKU? Sickle-cell anemia is at least 30-fold more common than PKU, and it is potentially just as devastating. Why hasn't sickle-cell anemia been eliminated from the gene pool, since natural selection acts so strongly against those who have the disease?

The reason why sickle-cell anemia remains a common problem for some ethnic groups is that there is a subtle and perhaps unusual interaction between the environment and the sickle-cell gene. Sickle-cell disease is most common in those parts of the world where malaria is a major killer. Malaria afflicts between 250 and 300 million people in parts of Africa, Latin America, South America, and Asia, and it causes more than a million deaths each year. But, for some reason, people who have the sickle-cell trait are more resistant to the blood parasite that causes malaria. In other words, those people who are heterozygous for sickle-cell disease actually have a survival advantage over those who are homozygous for the normal hemoglobin gene. This situation is called "balancing selection" since selection is working in two different directions at the same time. The result is an odd compromise between elimination of the sickle-cell gene and elimination of the normal hemoglobin gene.

Very recently, scientists found evidence that there are other examples of balancing selection and other remarkable ways in which genes and the environment can interact. Cystic fibrosis

(CF) is approximately ten fold more common than PKU, and it causes a devastating syndrome of lung infection, digestive tract problems, and malnutrition. Until relatively recently, the life expectancy of a child with CF was about 1 year. Since natural selection acts so harshly against a child with CF, it is hard to explain why the mutation has not completely died out. Yet very recently it was found that mice heterozygous for the CF trait are more resistant to cholera.[5] Cholera is an infectious disease that is still epidemic in many parts of the world, and it was a major killer in the United States until about the turn of the century. If human carriers of the CF gene are also more resistant to cholera, then this may be another example of balancing selection at work. Thus, selection would work to maintain the CF gene in the population, even though people who develop CF have a very much shorter life expectancy.

One can only imagine the range of interactions possible between the environment and other more complex multigenic human traits. For example, intelligence is known to be heritable, but heredity is often a poor predictor of intelligence. This is because the role of the environment is so very critical in determining that trait we call intelligence. A person blessed with genes that could confer intelligence will not actually be so unless he experiences a favorable or at very least benign environment. Prenatal exposure to alcohol, drugs, or disease could prevent a child from achieving the full measure of intelligence possible from a genetic standpoint. Similarly, adequate nutrition and freedom from physical injury and abuse are critical for the growth of intelligence, from infancy through adulthood. An environment rich in parental care and attention is obviously critical for survival, as King Frederick II proved, but it is also necessary for adequate intellectual growth. In addition, a rich stew of environmental stimuli is probably necessary to challenge the child and to optimize intellectual capacity. Finally, exposure to proper schooling is necessary throughout childhood and adolescence if a child is ever to achieve adulthood with a modicum of what is recognized as intelligence. To further complicate the picture, it may be that all of these things are required to occur in a very proscribed and specific way. For example, it may be that adequate nutrition is especially critical between the ages of 3 and 6 months if a child is to develop fully those intellectual skills that have their genesis at this time. The interactions between gene and environment, involved in unfolding the latent mind of an infant, are something completely beyond the ken of our current science.

The Synthesis of Nature and Nurture

During the development of a human embryo, more than 100 billion fledgling nerve cells send out long shoots called axons, which form an estimated 100 trillion connections with other cells in the nervous system. These connections are called synapses, and synapses between nerve cells are the building blocks of intelligence. Many axons have to travel long distances before they build synapses, and all synapses involve extremely precise connections between various nerve cells. The fact that this can happen at all is a miracle. The fact that it happens in much the same way, generation after generation after generation, is genetics.

Genetics determines the pattern and the schedule of human development, from conception and birth through adolescence to adulthood. Genetics even determines the pattern and schedule of aging. What genetics cannot control are the external events that occur during the internal processes of growth and development. Without a doubt, external events can have a profound effect on internal processes, but internal processes can also have a reciprocal effect on external events. In a crude sense, we cannot do what we are not programmed to do. But equally true, we cannot do what we are programmed to do if the environment does not facilitate so doing. Environment can amplify or blunt the effect of genes, but environment cannot replace or displace genes.

After the connections between neurons are made during development, it is up to the organism to use those connections.

Use may strengthen some connections or lack of use may dull others, but use cannot make new synapses unless the genes so specify. Once a synapse is established, it is subject to a sort of natural selection; either it is used or not, in a way that is at least partially determined by available stimuli. This concept of neurons forming synapses, whose activity is then modulated by the environment, owes much to the work of Donald O. Hebb of McGill University. Twenty years ago, long before there was much solid evidence in support of the idea, Hebb taught that "heredity determines the range through which environment can modify the individual." The development of molecular biology over the last 20 years has done much to confirm Hebb's ideas.

2

The Old Nature versus Nurture Debate

The debate about whether nature or nurture is more important in determining human behavior is an old one, but every so often it erupts anew into a bitter and acrimonious fight. Most recently this was shown by the virulent controversy surrounding publication of *The Bell Curve,* but this is not the first, nor will it be the last, such controversy. Usually those who argue loudest and longest do so from an ideologue's viewpoint, and few scientists have wanted to become embroiled in an argument that is so often divisive. Some individuals seem to see humans as automatons whose every action is controlled by genes, irrespective of what choices the environment presents. Others seem to see nature as constantly at war with nurture for control of the individual, giving the phrase "nature versus nurture" a new meaning entirely. Still others see humans as a *tabula rasa,* or blank slate, written upon by experience; simply give a child the right set of experiences and that child can grow up to be an athlete, an artist, or a mathematician. And the pendulum of public opinion has swung back and forth, sometimes deifying and sometimes villifying scientists who study the origins of human behavior.

A great deal of confusion surrounds the subject of human behavior, and this confusion is not the fault of the lay public; when dealing with issues as complicated as behavior, confusion is inevitable, and arises for many reasons. Behavioral genetics is a brand new science, which has so far provided us with many more questions than answers, and genuine confusion exists within the field. Just as most movies, books, and cars do their job in a pedestrian and prosaic fashion, most scientific studies are also workmanlike but undistinguished.[6] Virtually never is a study so well done that it is accepted as the last word on a subject. Sincere scientists, acting in good faith, can report data that later prove to be incorrect, while other scientists may attempt ambitious studies that ultimately fail to obtain a clear-cut answer to any question. Overcautious scientists may be reluctant to report their findings, while other reckless scientists may report preliminary findings from small studies as though these findings were established fact. Members of the media are often gullible enough to accept preliminary findings at face value, failing to see the problems that ultimately undermine a poorly done study. Many newspapers and television news shows are guilty of presenting an oversimplified summary of a complicated issue, in order to fit the story into a particular newspaper space or bit of air-time. And some writers about science have willfully distorted or misrepresented the truth in service to a political ideology. The lay person is bombarded with utterly fallacious pseudoscience while standing in the grocery checkout line, and many stories presented in respectable newspapers are incomplete, distorted, or incorrect. Most people simply don't have the time, energy, or inclination to separate grains of scientific truth from the chaff of misinformation. But these various problems have not stopped the average person from forming an opinion.

Through most of the history of biology as a science, there has been a subtle but pervasive bias that nature, in the form of hereditary forces at work in the individual, is dominant in the origin of animal traits. Often, by extension, human traits are also seen to result from the unfurling of an immutable program borne in the genes.[7] The bias toward genetic determinism may have originated with animal breeders, who selected for specific temperaments as well as specific physical traits in domestic animals. But the bias was blessed by science in the era following Darwin and Galton.

The "Split Twin" Approach to Separating Nature and Nurture

The first scientific confirmation of the importance of genetics in human behavior came from the study of twins who had been separated at birth and reared in different households, so that they experienced completely different environments. This kind of "split twin experiment" is able to address the importance of both nature and nurture, since the effects of genes and the environment can be separated. This type of experiment is best when it involves both identical and fraternal twins, since this means that nature conspires with the scientist to reveal the importance of the genes. Identical twins are derived from a single egg, which splits into two separate embryos, so identical twins share identical genes. In the psychological literature, identical twins are usually referred to as monozygotic (MZ) twins, which relates to the fact that these twins are derived from a single fertilized egg. Fraternal twins, often called dizygotic (DZ) twins, are instead derived from two separate fertilized eggs. Since the two separate eggs must be fertilized by two separate sperm cells, the genetic differences between fraternal twins can be extensive. By having two different types of twins with two different degrees of relatedness involved in the same experiment, it is that much easier to see the relative importance of genes and the environment.

It is important that a split twin study should include as many twins as possible. This is because there is always some degree of randomness and uncertainty in the expression or in the measurement of human traits, and this randomness can only be factored out if a study includes many subjects. It is simply not adequate to describe strange similarities between separated twins; however striking these coincidences, the evidence is not worth reporting from a scientific standpoint. It is thus very difficult to do split twin research because it is usually very difficult to identify an adequate number of twins split at birth.

Only by having access to information on millions of births, through a central registry of birth data, is it possible to accrue the hundreds of subjects necessary to make a strong "split twin" study.

If twins split at birth are compared to twins reared together, this provides insight into the role of the environment in the genesis of human traits. For example, if identical twins reared together have very similar intelligence, whereas identical twins reared apart have different intelligences, this would imply that the environment is critical in determining intelligence. And if both identical and fraternal twins reared in the same environment are very similar in intelligence, even though they are not genetically that similar, this would be a convincing argument that intelligence is largely molded by the shared environment of the home. Alternatively, if identical twins reared apart are as similar to each other as identical twins reared together, this would argue that the environment is much less important in determining intelligence.

If genetics is central in determining human traits, then identical twins should be much more similar to each other than fraternal twins reared under the same circumstances. In fact, this is generally found to be true, and should not be too surprising. If identical twins are reared together, many parents feel obligated to give them similar names, to dress them the same, to treat them the same, and to present them with the same opportunities. This is perhaps less often true of fraternal twins reared together. However, if identical twins are reared apart, they are still more likely to be similar to each other than fraternal twins reared apart. This is because a child's experience of the world is very strongly influenced by whether that child is male or female; identical twins are always of the same sex, whereas fraternal twins are of opposite sexes nearly half the time.

The first description of a split twin experiment was reported in 1869 by Francis Galton, who also studied the heritability of genius. Since then, countless scientists have taken this approach to study the relative contribution of genes and the environment to virtually every human trait imaginable. In fact, split twin studies remain as one of the two pillars of behavioral genetics, with molecular genetics forming a newer but no less substantial pillar. Almost invariably, the old split twin experiments reported that a relatively high proportion of human variation is inherited.

Some of the more modern split twin experiments have confirmed the genetic determinism of earlier studies.[8] For example, the Minnesota Study of Twins Reared Apart found that identical twins reared apart were very nearly as similar to each other as identical twins reared together. Variations in intelligence were found to be about 70% related to heredity, meaning that education, social interaction, nutrition, and a generally supportive environment together are only half as important as genes in the development of intelligence. Traits such as personality, temperament, occupation, leisure-time activities, and social attitudes were also very consistent between identical twins. Stories of eerie parallels between identical twins reared apart were frequently reported in popular press descriptions of this study. One set of identical twins included Oskar, raised as a Nazi in Czechoslovakia, and Jack, raised as a Jew in Trinidad. When they were reunited, both men were reportedly wearing shirts with epaulets, both flushed the toilet before and after using it, and both enjoyed sneezing suddenly in elevators to startle other passengers. Two reunited female twins were reported to be wearing seven rings each and to have named their firstborn sons either Richard Andrew or Andrew Richard. Two male twins were both named Jim, had married women named Linda, had divorced and remarried women named Betty, and had pet dogs named "Toy." Some of these similarities between separated twins are so bizarre or so eccentric that it is hard to imagine two people doing such strangely similar things merely by chance.

This study, and others like it, led to the pervasive modern attitude that all traits are inherited. This modern attitude is reinforced by a popular bias toward the "hard sciences" like molecular biology, because the results of these sciences are supposed to be less ambiguous than the "soft sciences" or humanities. In fact, there is often an explicit bias against the "soft sciences" like psychology, because these sciences can so seldom be definitive. But bias aside, the very latest results imply that the old split twin studies tended to overestimate the importance of genetics. The reason for this overestimation is that there are many problems and difficulties associated with doing a proper split twin study.

Split Twin Studies Are Fraught with Major Problems

A major problem with split twin research is that it is so difficult to understand the influence of the environment in a meaningful way. Any study of twins, whether identical or fraternal, must confront the fact that similarities in the environment of two individuals can arise in many different ways. Two strangers, born on the same day in the same place, will likely share a certain portion of the environment. And, while young children do not experience much outside the home, their parents do, and parents can bring external influences into the home. Any child born in the United States in the years after World War II was exposed to the pervasive paranoia of the Cold War. To a certain extent, the vaguely sensed anxiety of our parents, the bomb shelters and air raid drills, the press coverage of the Red Menace, and the general frenzy following the Russian launch of Sputnik, weave all Baby Boomers together in a shared experience of the environment. Similarly, the Great Depression was an experience that impacted virtually every household in the early 1930s, and may have left a mark on virtually every child of that era. Today, many children share the experience of poverty and deprivation, and this experience is likely to leave a profound mark as well. People of different backgrounds but similar ages can share many experiences, and may develop many similar personality traits in response to those shared experiences.

Another major problem with split twin research is that many of the traits that are of greatest interest, such as intelligence or proneness to violence, are very difficult to measure in an objective way. All of the early psychological measurement tools were primitive, and many of the tools in use today are probably still primitive. If a test that measures some psychological variable is not accurate and objective, then any measured similarity between two people is basically meaningless. Because psychological tests have evolved rapidly, it is essentially impossible to compare current generations to past generations, because the different generations were likely given different tests. Comparing the results of two different tests, even if they were intended to measure the same trait, is much like comparing Porsches to Pontiacs. And to further compound these difficulties, it is often very difficult to get scientists to agree on what constitutes an amorphous trait like intelligence.

The absence of good data over time makes it very difficult to characterize the heritability of a trait, since the data on a given trait may go back for only one generation. In fact, our inability to get accurate information on past generations is quite pervasive. Family histories are notoriously unreliable, so it can be very difficult to identify traits of interest in past generations. The claim that old Aunt Jean remembers her Great-Uncle Arnold as having a medical condition suspiciously like schizophrenia is almost certainly unreliable. In the past, the medical sophistication of most people was insufficient to make such a diagnosis, and memory is a mutable thing anyway. A family mythology can grow up around an individual no longer present to defend himself. It is quite likely that a large number of pathologies in the past were overlooked because of the lack of medical knowledge, or these pathologies may have been forgotten with the passage of time.

Another major problem with split twin studies is that there is very often bias in the selection of study subjects. Recruitment methods used to gather subjects for a study usually do not eliminate the possibility of self-selection by subjects. Split twins who volunteer to participate in a study are probably more likely to have maintained contact with one another than usual. If this is true, then they are probably also more likely to be similar to one another. It is best to recruit twins to a study using a registry that records the birth of all twins, but this is often not possible. In the absence of such a twin registry, the scientist may be dependent on one twin knowing how to get in touch with the other twin. There must then be a suspicion that the actual similarity between split twins is somewhat less than the measured similarity between split twins.

When a large number of identical and fraternal twins are studied, the data are often analyzed to determine the "heritability" of a particular trait, or the extent to which heredity controls the expression of that trait. However, there are a number of problems with this approach. Perhaps the biggest problem is that the "heritability" calculated for a trait is only accurate for those people from whom data were drawn in the first place. As an example, let us assume that weight is 70% heritable; the average weight of a group of people in New York is 70% heritable, as is the average weight of a group of subsistence farmers in Somalia. Nevertheless, we cannot assume that the average weight difference between an overfed New Yorker and an underfed Somalian is caused by genes. Clearly, genetic differences between these groups of people would exist, but the average difference in weight is far more likely to be caused by the shortage of food in Somalia. This means that every time a scientist calculates heritability in a group of white male twins and then projects that degree of heritability to another population, even a population of white male nontwins, the results must be at least somewhat suspect. When the heritability calculated from a group of white male twins is projected to black male and female nontwins, the results are likely to be substantially wrong. Yet this kind of thing is done all the time, even by highly reputable scientists.

"Heritability" is very hard to determine accurately, because the mathematical model of heritability makes a number of important assumptions that may or may not always be true.[9] For example, the model assumes that both identical and fraternal twins raised in the same household share precisely the same environment. Yet this is unlikely to be true, because parents seem to treat identical twins differently than they treat fraternal twins. In fact, identical twins may experience a much more similar environment than do fraternal twins, so that the effect of the environment would differ for identical and fraternal twins. The mathematical model used to calculate heritability may thus fail specifically in comparing identical to fraternal twins, which would be fatal to calculations of heritability.

Additional Problems Are Common to Most Behavioral Genetics Studies

In addition to the problems discussed above, which are really unique to split twin studies, there are a number of other more general problems that can wind up sinking any behavioral genetics study. For example, the mathematical model used to calculate heritability of a trait can fail because some very general assumptions are violated, and these assumptions may be common to molecular genetics studies as well. The model used to calculate heritability assumes that all genes act additively, so that the individual genes simply sum up to some larger effect. In other words, the effect of one gene is assumed to increase the effect of another gene only; no synergy between genes is permitted, nor is it possible to have one gene cancel the effect of another gene. Yet, in real life, genes may interact in all kinds of subtle ways that the model does not acknowledge. If the model assumes that a synergistic interaction does not occur, yet that interaction occurs anyway in real life, then the model will calculate an incorrect "heritability" for the trait in question. For example, assume that there are two genes that can produce obesity; perhaps one gene dulls the sensation of satiety, while another gene affects metabolism such that fat storage is more likely. A person inheriting either of these genes might have a tendency to become somewhat obese, but a person inheriting both genes might have a strong tendency to morbid obesity.

Even more problematic, the mathematical model used to calculate heritability assumes that there is no such thing as "assortative mating."[10] This means that the model requires people to mate at random, so that children are a random mixing of maternal genes and all possible paternal genes. Yet, in real life, assortative mating happens all the time. Women for whom health is important probably try to select a healthy mate, whereas an alcoholic woman may select a man who shares her problem. Intelligent women are more likely to bear the children of a man who is also intelligent, while women who are not intelligent may be unable to attract an intelligent man. If a woman is very tall, this may mean that all possible mates are subconsciously screened for similar height. Yet, for the model to work well, there can be no screening for height or for any other trait; mating must be entirely random. Because assumptions inherent in the model are violated, the model itself will fail; this means that, even with perfect data (and data are never perfect), the heritability of a trait will be incorrectly calculated. If mating is not random (as it surely is not), then the genetic component of behavior would be underestimated, since part of the "environment" is actually created by selective mating.

A final problem with much of behavioral genetics research is that there has been a general tendency among scientists in the field to use statistics naively or badly, and the whole field suffers from a general weakness of statistical methods. All of science depends on making careful and precise measurements many, many times, so that eventually one has confidence that the numbers one measures are correct. Unless a measurement has been made many times, a scientist simply does not know whether the "true and correct" answer has been obtained; it is not possible to check the answers in the back of the book when the book has not been written. Yet it can be so difficult to make the measurements necessary for a behavioral genetics study that there is a constant temptation to stop collecting data too soon. At least one moderately large molecular genetics study, which examined the inheritance of manic depression,[11] was completely undone by the fact that the study included too few subjects. If too few subjects are analyzed, the strength of a study may depend on the diagnosis of one or a few subjects; in fact, the manic depression study was undone by only two missed diagnoses. In this study, two subjects who were initially diagnosed as well and healthy later developed manic depression. When the diagnosis of these two subjects changed, the conclusions of the whole study had to be thrown out.

Far too often an inadequate number of people are analyzed in behavioral genetics research, so that a scientist cannot make any conclusion with confidence.[9] This can be particularly damaging in some cases, because careful analysis shows that it is easier to disprove a genetic model of inheritance than an environmental model of inheritance. This means, of course, that there are likely to be more errors when attributing a trait to the environment than when attributing a trait to genetics. A mathematical analysis has shown that, if a scientist starts with an incorrect model of how a trait is inherited, it can take more than 600 pairs of twins to disprove that model of inheritance. Generally speaking, a split twin study should be structured so that it involves more fraternal than identical twins, in order to maximize the chances of correctly determining inheritance. In fact, there should be roughly twice as many fraternal as identical twins in a study, in order to be sure that the calculated inheritance is the true inheritance.

According to a recently published analysis, it is easier to identify a linkage between a gene and a trait than it is to confirm that linkage in a subsequent study. To show this, scientists made an assumption that a particular trait is 50% heritable, meaning that 50% of the variation in the trait is explained by heredity. If that trait affects 10% of people and is associated with six different genes, then mathematical analysis shows that a scientist would need to enroll 2000 people in a study to detect the effect of one of those six genes. At least 175 separate families would need to be analyzed, assuming that each family has four grandparents, two parents, and four children. If any of the families was smaller (as most are), then an even larger number of families would need to be enrolled to be sure of detecting the effect of one gene. This is a huge number of subjects and an enormous amount of work. Yet, if a scientist is interested in confirming a finding by repeating a study, an even larger number of people is required. To confirm or "prove" that one gene among six played an important role in the inheritance of a common trait would require as many as 8000 new subjects. This would mean enrolling 781 new families just to confirm a result that has already been published. Few granting agencies are willing to finance such a study because the result of the study is already known. Yet the very nature of science requires that a finding or a result be replicated before it can be believed. It is just too easy for a well-meaning and dedicated scientist to make a mistake, and there are some scientists who are neither well-meaning nor dedicated. To accept an important result at face value, without verifying it, is sheer foolishness. Yet the budget constraints that science now operates under mean that many important results may never be replicated until too late.

Underestimating the Environment

This list of caveats is not meant to imply that split twin studies are invariably weak or that molecular genetics studies are inherently stronger. Split twin studies, when properly done, are an astonishing window into the workings of the human genetic program and how that program interacts with and is influenced by the environment. It is quite likely that split twin studies will remain as one of the two principal ways to attack the problem of nature versus nurture, with molecular genetics being the other approach. The caveats here are simply meant to illustrate the difficulty of doing science properly. And certainly molecular genetics has problems of its own, which will be examined in greater depth later (Chapter 5).

The truth of the matter is that genetics simply cannot explain all of human behavior. The most recent information suggests that roughly half of all that makes us human is inherited from our parents. This of course means that roughly half the range of behaviors we see around us are the result of the environment acting on the individual. Without a doubt, the environment and the genes interact in subtle and marvelous ways that have not yet been deciphered. But there is very grave danger in accepting that all behavior is genetic.

3

The Dark History of Eugenics

A large part of the reason why the subject of nature and nurture is so laden with emotion for so many people is that the new science of behavioral genetics has intellectual roots in the old ideas of eugenics. Eugenics is that field of study dealing with improving the inborn qualities of the human race, particularly through the control of hereditary factors. The emotional resonance of the nature versus nurture controversy cannot be fully appreciated without an understanding of the dark history of eugenics. No one should forget that the ideas touched on here are explosive; problems in human biology are fascinating, but they are also emotionally charged. It is simply impossible to study our own species as dispassionately as we would study an insect or a bird. Social values are inherent, or potentially so, in any scientific finding about humans, and the scientist who is unaware of this is naive and open to exploitation.

Eugenics originated as a scientific movement, validated by the leading scientists of the time. To call eugenics a "pseudoscience" is to make it seem less threatening, but it is also incorrect; the great majority of scientists at the turn of the century believed in eugenics. In 1916, all five scientists who founded the American journal *Genetics* were advocates of eugenics, even though each was an established scientist of great reputation.[12] If most practicing scientists adhere to a certain view of the world, that viewpoint is, by definition, mainstream science.

Although eugenics began as a scientific concern for the betterment of the human race, it evolved into a social and political effort to control human evolution. The ideas of eugenics were gradually perverted into ideals, against which all persons could be measured. Ultimately, millions of people were systematically killed by the Nazis because they did not fit the rigidly codified ideals of the day. The fact that mainstream science was used as a rationale for systematic genocide is proof that ideas can have great power. It is also proof that scientists who have ideas of great power may be unable to foresee the consequences of those ideas.

The word *eugenics* was coined in 1883 by Francis Galton, who, as mentioned earlier, first achieved prominence studying the heritability of genius. The word is derived from Greek words meaning "wellborn" and it refers to the application of scientific principles to improve the human stock, by ensuring the production of healthy offspring. A eugenic program is thus a public policy designed to increase the frequency of some desired trait in a population. A eugenic program can involve positive eugenics, or a systematic effort to maximize the transmission, from one generation to the next, of traits that are considered desirable. But a eugenic program can also involve negative eugenics, or a systematic effort to minimize the passage of traits considered undesirable. It seems to be a relatively small intellectual step from a program of negative eugenics to one of genocide, or the systematic extermination of persons who possess traits considered undesirable.

The Early History of Eugenics

The intellectual roots of eugenics extend all the way back to Plato, who believed that defective children should not be cared for by parents. He also believed that chronic invalids and those who were ill because of self-indulgence should not receive medical care, and that moral degenerates should be executed. Plato even advocated temporary unions between superior men and women for the express purpose of having superior children.

The rationale for a eugenics program was clarified in 1798, when Thomas Malthus proposed that the human population was expanding faster than was the food supply. His *Essay on the Principle of Population* was the first to note that the earth's resources are limited, and that starvation awaits a population that outgrows its resources. Malthus believed that increases in population would always outstrip increases in food production, and that rampant population growth would eventually be checked by competition between human beings for the simple necessities of life. This essay is often credited as the first substantive contribution to the study of ecology, but it is also arguably the first major contribution to both evolution and eugenics. Darwin was strongly influenced by Malthus, and Malthusian ideas of resource limitation are a clear precedent to the idea of natural selection. When Charles Darwin published his *Origin of Species* in 1859, all the intellectual tools were then in hand for the development of eugenics.

Francis Galton was convinced, as long ago as 1861, that a wide range of human physical, mental, and moral traits are inherited.[13] Galton had read *Origin of Species*, written by his cousin Darwin, and it made a great impression on him. Galton reasoned that continued progress or evolution of humans depended on transmitting the best human traits to future generations. In 1865, Galton proposed that human society could be improved through "better breeding" in an article entitled "Hereditary talent and character." But Galton's ideas on eugenics caught on slowly because, at that time, there was no understanding of genetics, or the mode of transmission of traits from one generation to the next. Mendel's now famous experiments on the breeding of pea plants had already been done and forgotten, and Western science had not yet rediscovered Mendel's laws of inheritance. But, in 1883, Galton published his book *Inquiries into Human Faculty*, in which he first used the word *eugenics* and described in detail his ideas for improving the human race by controlled breeding. This time his ideas fell on fertile ground, and the Galton Laboratory was soon founded at University College, London.

Soon after Galton wrote his treatise on the heritability of genius, another highly influential book was published, this one on the heritability of criminality. *The Jukes: A Study in Crime, Pauperism, Disease, and Heredity* was about a family who came to the attention of the New York State judicial system because six members of the family were in prison in just one county. The book, published in 1877, included a detailed genealogy of the Jukes family, covering seven generations and including 750 different members of the family.[1] The author contended that the Jukes were an example of an inferior bloodline, one that was essentially doomed to poverty, vice, and crime, and that this family had already cost the state of New York well over $1.5 million.

Eugenics in Germany

Mendel's work with pea plants was rediscovered in Germany at the turn of the century, and it inspired a tremendous flurry of scientific research in genetics. Germany quickly became the European center of activity in genetics, and it also became the center of activity in eugenics. Mendel's laws of inheritance were soon invoked to explain many different familial patterns of inheritance, including the inheritance of mental illness, retardation, alcoholism, criminality, prostitution, and poverty.

Around the turn of the century in Germany, a wide general interest in "racial purity" was stimulated by two things: social unrest caused by rapid industrialization and concern about the societal implications of Darwin's theory.[13] It was generally believed that elite segments of the German population were producing too few children in comparison with the working classes. The various wars that Germany had been in were seen as counterproductive, since the most able men were sent off to die, while those who had been rejected for military service stayed home and were free to procreate. The German physician Wilhelm Schallmeyer proposed in 1892 that each citizen be given an annual examination by a physician trained in the science of heredity. The latter was to be a state official and each German would be issued a health passport. Another young physician, Alfred Ploetz, returned to Germany from the United States in 1895, to publish a book on "racial hygiene." Schallmeyer and Ploetz were quickly acknowledged as the intellectual leaders of the eugenics movement in Germany.

Perhaps originally, the eugenics movement in Germany was concerned with the betterment of the German nation as a whole, but this was quickly perverted into concerns for the betterment of "Aryans" alone. Aryans were rigidly defined as those non-Jews who could document their German origins for many generations back. National fitness was linked to "racial fitness," and the supposed inferiority of Jews, Eastern Europeans, and blacks became a constant theme, as government regulation became more and more intrusive. In 1908, in the German colony of Southwest Africa, all existing interracial marriages were summarily annulled and outlawed. In 1913, the prominent anthropologist Eugen Fischer called for a national network of clinics for the genetic screening of the entire German population. There was soon an effort to set up marriage-advice clinics to screen those about to be married, in order to establish hereditary health. By 1928, there were 224 marriage-advice clinics in Prussia alone.[14] Little attention was paid to confidentiality or to individual rights, and human health was valued as a national resource, rather than as an individual blessing.

By 1929, Adolph Hitler felt free to speak in public about killing German infants with physical defects, at a rate he estimated to be up to 700,000 children per year.[13] In 1933 the Nazi party came to power, and they began more actively to promulgate their ideas on racial purity. Physicians, who formed the largest professional group in the Nazi party, were encouraged to see themselves not as doctors to the individual, but as doctors to the nation. In 1934 an Office of Racial Policy was established by the Third Reich, to enlighten the public as to the benefits of applied racial hygiene.[15] The first action of the Office was to establish a nationwide system of genetic health courts, and physicians were required to register all cases of genetic illness with these courts. A sterilization law was passed which enabled the state to force sterilization on the infirm, and as many as 400,000 people were involuntarily sterilized. The principal reasons given for forcible sterilization were congenital feeblemindedness, schizophrenia, or hereditary epilepsy.

Within 10 years Hitler's plan to euthanize children also became reality; the Committee for the Scientific Treatment of Severe Genetically Determined Illness was established in 1939, with a mandate to destroy retarded and deformed children. Under the direction of this committee, forms were sent to physicians throughout Germany, ordering them to register deformed or retarded infants with a central authority. Children selected as having "lives unworthy to be lived" were selected and sent to one of 28 killing facilities.[13] At these facilities, either basic care was withheld until the children died, or else the children were deliberately killed. An estimated 5000 children were killed in this program, usually by morphine injection or cyanide poisoning. Standardized letters were then sent to parents informing them that their child had died an unexpected death because of health problems. This program began by exterminating children up to the age of 3 years, but by 1941 the program had been extended to include those up to the age of 17 years.

Hitler also authorized an adult euthanasia program and ordered that certain doctors be given the power to confer a "mercy death to patients judged incurably sick." In the infamous "T4" program, institutionalized mental patients were targeted for death; six facilities were set up to kill the mentally ill, using gas chambers disguised as showers. A total of 70,273 people were executed in this program, as detailed in meticulously kept official records.[15]

But this was only the beginning. Eugenics and the idea of racial purity were soon used as a rationale for the Holocaust, in which perhaps 10 million Jews, Gypsies, and other "undesirable elements" were slaughtered. Jews specifically were targeted for extermination, to improve the purity of the Aryan race. But the SS did not wish to kill all Jews outright; they wanted a ready source of slave labor as well. Healthy and robust Jews were selected for slave labor, and breeding experiments were begun in order to maintain a steady supply of slaves. But, to ensure that the reproduction of Jews could be controlled, experiments were conducted in the death camps, to determine how rapidly men and women could be sterilized, usually without benefit of anesthesia.

The Hitler Youth and the SS were also part of a massive Nazi program of controlled eugenics. An agency of the SS screened all SS candidates and their fiancées, to determine if these individuals were sufficiently Aryan. Applicants were required to produce documentation of Aryan ancestry going back as far as 1800.[13] German women generally were encouraged to have children out of wedlock with SS officers, as it was an honor to bear children for the Reich. Arranged marriages were common among members of the Hitler Youth, based on the purity of bloodlines and the possession of health, intelligence, and an Aryan appearance. Another branch of the SS deliberated on the qualities of children from conquered nations, and determined the potential of these children for "Germanization." Thousands of children were summarily taken from their parents in the Netherlands, so that these children could be indoctrinated as Nazis.

Eugenics in the United States

Most Americans know that a great evil was unleashed in Europe, particularly in Germany, in the name of a eugenic ideal. But most do not know that we as a nation are also guilty of abuses in the name of eugenics. Our country has taken more than one step down the slippery slope that leads from eugenic ideas to eugenic ideals.

Interest in eugenics was widespread in the United States at the turn of the century, because many believed that insanity, poverty, delinquency, and criminality were hereditary. As long ago as 1897, a eugenic sterilization bill was introduced into the

Michigan state legislature, calling for castration of the feeble-minded and of some criminals.[13] Ultimately the bill was defeated, but several institutions proceeded with eugenic sterilization anyway. At the Kansas State Institution for the Feeble Minded, one doctor sterilized 44 boys and 14 girls before public outcry stopped him. In 1907, Indiana became the first state to pass an involuntary sterilization law based on eugenic principles. This law required sterilization of all inmates at state institutions who were insane, feebleminded, convicted of rape, or habitual criminals. Forced sterilization was legal in 30 states by 1931, and the laws applied to an ambiguously wide range of "defectives," including "sexual perverts, drug fiends, drunkards, epileptics, and diseased degenerate persons." In 1961, 28 states still had laws allowing sterilization of the hopelessly insane or the feebleminded. The degree to which these laws were enforced varied greatly, but 12 states (California, Georgia, Indiana, Iowa, Kansas, Michigan, Minnesota, North Carolina, North Dakota, Oregon, Virginia, and Wisconsin) each performed more than 1000 sterilizations. As of 1961, there had been 62,162 sterilizations in the United States, with roughly equal numbers of sterilizations performed on the mentally ill and on the feebleminded. Recently, there has been a trend to repeal sterilization laws but, as of 1987, eugenic sterilization of institutionalized persons was still legal in 19 states, even though the laws are seldom used.

Many states also passed miscegenation laws, meant to prevent interbreeding between races, or what was referred to as "racial mixing."[12] Most miscegenation laws were meant specifically to prevent unions between whites and blacks, based on the flawed assumption that children born to these unions would be inferior. In fact, much of the rationale against cross-racial childbearing was derived from experiments in which two breeds of animal were crossed. Edward East, a prominent Harvard geneticist, summarized the reasons for prohibiting interracial crosses when he said it was "an illogical extension of altruism . . . to seek to elevate the black race at the cost of lowering the white" because "in reality the negro is inferior to the white. This is not hypothesis or supposition; it is a crude statement of actual fact."[16]

The eugenics movement in the United States was firmly established by 1910, when the biologist Charles Davenport set up the Eugenics Record Office in New York.[13] Davenport was a prominent scientist who had made his reputation working out the genetics of Huntington's chorea, as well as demonstrating the heritability of eye, skin, and hair color. His Eugenics Record Office was meant to carry out research in human heredity, especially the inheritance of behavioral traits, as well as to educate the lay public about the implications of eugenics for public policy. This office gathered detailed records on thousands of families, and used genealogical information to argue that various social ills, including criminality and poverty, were genetically based. Davenport even argued in 1919 that the ability to be a naval officier was inherited from two genes: a "thalasso-philia gene," for love of the sea, and a "hyperkineticism gene," for wanderlust.[17] Davenport believed that the paucity of female naval officers proved that these genes were unique to men. This is a perfect example of the naivete with which claims of herita-bility were made and accepted in that era. Davenport was apparently so blinded by science that he never considered the possibility that there were few female naval officers because of an active effort to keep women out of the navy, and especially, to keep them out of the officer ranks.

The eugenics movement in the United States was given respectability and popular appeal by a book written in 1916, entitled *The Passing of the Great Race*. In this book, which might as well have been called *In Praise of Anglo-Saxons*, the author made the anti-Semitic statement that "the cross between any of the three European races and a Jew is a Jew." Virtually every textbook on genetics written in the United States between 1910 and 1930 advocated eugenics in one form or another.[18] In one widely used textbook, *The Principles of Genetics*, which was published in the United States in 1925, the author states: "it is to be feared that even under the most favorable surroundings there would still be a great many individuals who are always on the border line of self-supporting existence and whose contribution to society is so small that the elimination of their stock would be beneficial." In fact, it is nearly impossible to find an American scientist expressing any opposition to eugenics in print until the mid to late 1920s.

An extremely popular book was published in 1921, which described yet another family as illustrative of a tainted bloodline. It told of a soldier who had fathered two lines of offspring, one a degenerate line that originated with a tavern maid, and the other a more respectable line born of a woman the soldier married. Over time, the degenerate line supposedly produced many children who were mentally retarded, while the respectable line produced no retarded children. The conclusion, of course, was that these intellectual differences were hereditary, but there was also the implication that virtue and morality are also heritable.

Beginning in the 1920s, the American Eugenics Society, founded by Charles Davenport and others, began to sponsor "Fitter Families Contests."[17] Whole families were judged like cattle at state fairs around the United States. An amusing photograph, of entrants in a "Fitter Families Contest" at the 1925 Texas State Fair, shows a family, graded in size from the father to the mother on down through five girls, all posed in identical bathing suits. From father to youngest daughter, each member of the family looks like a somewhat smaller version of the person to their right, and each is in a pose of simulated action. The only exception is the youngest daughter, who is looking at the camera in frank bewilderment. One wonders what criteria were used to judge fitter families; were families favored if they all looked the same, or were there more substantive criteria?

The United States maintained an open-door policy of immigration, but this began to change in 1921, when a temporary law was passed restricting entry to those who were financially well-off. This may have been done in an effort to reverse the unemployment and economic slowdown that occurred at the end of the First World War.[13] The Immigration Restriction League pressured Congress to pass permanent laws to limit immigration to those of Anglo-Saxon or Nordic descent. Calvin Coolidge, who perhaps should have stayed even more silent, said that "biological laws show ... that Nordics deteriorate when mixed with other races." That bastion of egalitarian thinking and liberalism, *The Boston Globe*, ran an editorial in 1921 entitled "Danger that world scum will demoralize America," while the president of Stanford University wrote that the "lower races," emigrating into the United States from Eastern Europe and Asia, would reduce "our own average."[12] After much debate in Congress, it was agreed to base future immigration on the mix of nationalities recorded in the 1890 census. The result was a piece of legislation called the Johnson Immigration Restriction Act of 1924, which severely limited immigration of peoples from Eastern or Southern Europe, especially those of Jewish descent.

Eugenics was often a thinly disguised rationale for racism of the worst kind. Edward East, a prominent Harvard geneticist who was a pioneer in hybrid corn research and one of the first to characterize multigenic inheritance, was also a virulent racist.[16] East believed that children born of interracial unions were likely to be genetically inferior to either of their parents because genetic crossing would "break apart those ... physical and mental qualities which have established a smoothly operating whole in each race by hundreds of generations of natural selection." East believed, on the basis of little or no evidence, that crosses between the races would produce disharmonious results, both in terms of physical traits and in terms of mental abilities. He was also one of the intellectual leaders of the fight to "cut off defective germ-plasm," or staunch the flow of supposedly inferior immigrants from Eastern Europe. This movement culminated in the Immigration Restriction Act of 1924. But East was not the only geneticist who clung to prejudiced ideas well into the 20th century. As late as 1931, *Human Heredity,* the most widely used and well-respected textbook in human genetics, made very sweeping claims about the racial and ethnic characteristics of people without losing credibility.[12] In various places throughout the text can be found the following quotes:

> Fraud and the use of insulting language are commoner among Jews. . . .
>
> In general, a Negro is not inclined to work hard. . . .
>
> The Mongolian character . . . inclines to (ossification) in the traditional. . . .
>
> The Russians excel in suffering and in endurance. . . .
>
> In respect to mental gifts the Nordic race marches in the van of mankind. . .

And, as late as 1942, two prominent American medical scientists advocated euthanasia for retarded children.[15]

"Environmentalism" as a Reaction to the Holocaust

The intellectual tenor of the time began to change, in Europe and in the United States, during the Second World War, in response to the repugnant racial doctrines of the Nazis.[16] Few geneticists were willing to argue, as had the Nazis, that racial mixing was harmful. In fact, the United Nations Educational, Scientific, and Cultural Organization (UNESCO) came out with a statement on race in 1951, signed by 23 prominent geneticists. This statement contained several passages that made a distinct break with pre-War eugenics. For example, the UNESCO statement says that ". . . no biological justification exists for prohibiting intermarriage between persons of different races." The UNESCO statement goes further to say, "Available scientific knowledge provides no basis for believing that the groups of mankind differ in their innate capacity for intellectual and emotional development." This is, in essence, an endorsement of the ideas of "environmentalism," the doctrine that all differences between individuals and races are related to differences in environment and upbringing.

Environmentalism became the fashion, and everywhere intellectuals backed away from eugenics and even from human genetics. Scientific journals deleted the word *eugenics* from their title, and it became very difficult to publish anything on eugenics. The American Genetics Association changed its official credo, from the "improvement of plants, animals, and human racial stocks" to the "improvement of plants, animals, and human welfare." Environmentalism was reduced to the absurd by B. F. Skinner, who argued that human personality developed as a consequence of the rewards, punishments, sensory inputs, and responses of the developing infant. Skinner seemed to believe the newborn was a blank slate, on which anything at all could be written by parents, siblings, teachers, or friends. Skinner grew famous from his experiments with pigeons in a Skinner box, in which he trained them to do various bizarre tasks for a reward. But he also believed that people could be programmed to behave in predetermined ways, in such a way that would better fit them for a Utopian society. In fact, Skinner went so far as to subject his own child to extended stays in a Skinner box, thereby proving that extreme environmentalism is nearly as prone to perversion as is extreme eugenics.

For many years, there seemed to be a blind spot in the vision of geneticists: it was perfectly correct to speak of the heritability of simple traits in simple organisms, or even to speak of the heritability of disease in humans, but discussion of the heritability of human behavior was tacitly avoided. This trend continued until 1975, when the book *Sociobiology: The New Synthesis* was published by E. O. Wilson of Harvard. This book created a firestorm of controversy because it made the claim that much about human society was explainable in terms of evolutionary forces. Since evolution acts only on the individual, many believed that the implication of the book was racist: differences in the level of advancement of two societies were explainable by differences in the level of advancement of members of those societies.

The Present and Future of Eugenics

Today there appears to be a more balanced view of nature and nurture: a combination of genes and the environment are assumed to play a role in human behavior. Yet the truce between warring camps is somewhat illusory. It is more correct to say that sociologists and biologists have agreed to disagree; most sociologists display an appalling ignorance of basic biology, and most biologists couldn't care less about sociology. When the two camps collide, warfare is as likely now as in the past. This was shown as recently as 1992, when a conference on the genetics of violence, to be sponsored by the National Institutes of Health, was cancelled because of a vocal minority opposing it. Peter Breggin, of the Center for the Study of Psychiatry, was quoted as saying that behavioral genetics is "another way for a violent, racist society to say people's problems are their own fault, because they carry 'bad' genes." Even more recently, the Department of Health and Human Services attempted to initiate a program of research on violence. This initiative was intended to gather together the separate threads of research into violence, but it offended many who believed that this research was inherently racist.

Political correctness has now hamstrung the ability of sociologists and psychologists to attack problems of social relevance, while leaving molecular biologists free to approach such problems. Yet biologists often fail to consider the social consequences of their research, and to bear in mind the value of diversity, and the principles of tolerance, racial equality, and ethnic sensitivity. In fact, modern biologists are quite caught up in the successes of molecular genetics and may be contributing to an unbalanced view of the role of genetics in human behavior. The remarkable success of molecular genetics in explaining many biological and medical problems has created the expectation that molecular genetics will be able to explain, and eventually to solve, all of humanity's problems.

There is now a misconception that genetics explains all human variation, and that what genetics decrees is immutable.[12] James Watson, Nobel laureate and codiscoverer of the structure of DNA, is quoted as saying, "We used to think our fate was in our stars. Now we know, in large measure, our fate is in our genes." The Human Genome Project has been packaged and sold to the public as a way to find solutions to many of our social, as well as our medical, problems. One prominent scientist is quoted as saying that the sequence of human genes is what "defines a human being," while another has called DNA "the blueprint for life." The idea that genetics determines our future has inculcated a sort of genetic defeatism, the clearest example of which can be found in *The Bell Curve: Intelligence and Class Structure in American Life.* This book argues that intelligence is genetically determined and essentially unalterable, and that we must somehow find a "valued place in society" even for the dull and ignorant, since we cannot possibly change their lot in life.

An extreme form of reductionism is rampant among biologists; the common viewpoint is that all phenomena, even the human mind, can eventually be explained in terms of chemistry or even physics. Several prominent universities have recently established laboratories to research the mind–brain connection, making the tacit assumption that the human mind can now be explained in chemical terms. Every kind of biology other than molecular is devalued as merely "descriptive" when, in fact, molecular biology itself is simply another level of description. There is a tendency for biologists to assume that a social problem always has roots in genetics, and that somehow genetics can provide a solution to the problem.[12]

The door is now open to a resurgence of the eugenic ideas that led to such gross excesses in the past. We as a society have been reluctant to discuss eugenics openly, although we have already begun to implement policies that can be seen as eugenic in nature. The widespread availability of tests that can diagnose medical problems in the unborn has made negative eugenic selection possible, and even acceptable. Current abortion laws enable a woman to abort any fetus that has an undesirable trait, whether that trait is a proneness to disease or simply being female when a male child is wanted. Effort has been devoted to establishing medical screening tests for sickle-cell disease and other genetic traits, which will create information that can easily be used in a discriminatory fashion. There has been a call to institutionalize orphans in state-run facilities, where they will likely suffer second-rate foster care that may preclude a productive adulthood. Many of these orphans will probably come from disadvantaged families, so that the orphanages could become a repository for those children for whom society does not have a place. Prisons have already been built to warehouse adults for whom we cannot find another place. The death penalty is again legal in many states, and studies have shown that this final solution is more likely to be invoked when a criminal is a member of a minority group. Efforts to liberalize laws that allow compassionate death or "assisted suicide" are seen as contributing to the dignity of death, but these laws must be evaluated in a eugenic context. When people speak of relieving the burden of suffering of the sick, one must ask whose burden is actually being lifted? Is it that of the individual, or his family, or health care workers, or society at large? Any answer other than "the individual" potentially comes from tainted motives, and begins to sound suspiciously like a eugenics program in action.

4

The Nature of Nurture
Defining Environmental Influences

The old dichotomy of nature or nurture, first made by Galton more than a century ago, has been used so far without exploring it more deeply. But this dichotomy is misleading for several reasons, and it is now time for clarification. Most scientists believe that the dichotomy itself is incorrect, and that virtually every trait is mediated by both nature and nurture, working somehow in concert. Furthermore, it is quite outdated to speak of "nature or nurture" when we know that nature, as it was meant by Galton, is equivalent to genes in modern parlance. As has already been discussed; a gene is simply that part of the DNA molecule that specifies the construction of a particular protein. Environment, as Galton used it and as it has been used since, is something of a catchall; everything that is not genetic is, by definition, environmental. But this is intellectually unsatisfying; we must somehow be more explicit about the environmental influences that affect expression of a trait. Or, to put it another way, what is the nature of nurture?

Environmental Influences Broadly Defined

Nurture is often taken to mean the social environment that surrounds and protects the child from birth to independence. This would include early interactions with parents and siblings, as well as the more sporadic interactions with whatever members of the extended family happen to be around. Somewhat later the environment expands to include teachers and friends, and these parts of the social environment assume greater and greater importance with the passing years. Finally, those persons with whom an adolescent or young adult has lasting friendships or love relationships play an increasingly important role, whether those persons are of the same or of the opposite sex.

But considering environmental influences to be synonymous with social influences is really very narrow and restrictive. Instead, environmental influences should be broadly defined as anything and everything not explicitly in the genes. This opens the door to many factors that might otherwise be overlooked or undervalued, including factors in the physical environment. And it also opens the door to complex interactions between genes and the environment that are neither entirely genetic nor entirely environmental. While most of these influences on behavior are still speculative, the idea that the environment has a complex and subtle impact on the individual is really not at all speculative.

In fact, the environment can have a profound effect on the individual before the individual is even born. Environmental toxins or drugs can pass through the mother's body, to impact directly on the unborn child in many ways. An example of this is the drug thalidomide, which was taken by pregnant women 40 years ago to control morning sickness, but which had a devastating effect on fetal development. Alcohol as well can have a profound effect on the fetus, and fetal alcohol syndrome is known to be associated with mild mental retardation. The hormonal environment that the child encounters *in utero* is also potentially important, and maternal hormones may be able to modulate or alter the effect of the fetal genome. In addition, the nutritional status of the mother can have a profound impact on the development of the fetus, and developmental patterns established early in fetal life may take an entire lifetime to work out.

After birth, the newborn infant is extremely vulnerable to vagaries of the environment, and the physical needs of the newborn must be met in order to ensure survival. The nutritional history of the mother can be important if the mother is nursing, and the early nutritional history of the baby itself is obviously critical. But feeding and freedom from physical harm or abuse is not enough; the infant must be loved, touched, cleaned, kept warm and free of disease, and given adequate stimulation and rest. Only when all of these physical conditions are satisfactorily met can the child begin to fulfill its genetic potential. Since mere survival can occur under conditions that are far from optimal for growth and development, this implies that the environment will often hinder a child from fulfilling its genetic potential. This is why it is so critical that care be provided by someone sensitive to the needs of the child.

Direct interactions between the infant and the mother immediately after birth are extremely important, and the relationship with the mother remains important for the entire life of the child. The personality of the child and that of the mother can interact in a very important way to give this relationship a structure that may last a lifetime. Many neonatal behaviors seem designed to elicit warmth and nurturance from the mother. But if a child is somehow deficient in giving these signals, or if the mother is deficient in receiving these signals, this deficiency can have a major impact on the child. The ability to elicit or give nurturance may in fact be hereditary, so that the same thing that interferes with a child's ability to elicit nurturance may interfere with the mother's ability to give it. In fact, it is likely that the hereditary component of a social behavior such as warmth or empathy is first expressed in interactions with the maternal "environment."

While it is clear that maternal acceptance and nurturance of the infant is critical, the optimal development of a child may also require a father figure or an extended family. It is quite reasonable to suppose that infants are directly or indirectly sensitive to the social supports surrounding their mother. Rearing an infant can be extremely stressful, involving many sleepless nights, many disruptions of prior routine, and many new anxieties, and the new mother often has little time or energy to deal with her own feelings. The importance of an effective social support network around the mother at this time cannot be overestimated. It is quite likely that infants are sensitive to this support network in some way or another, since maternal anxiety can be transmitted to the infant by many different cues. In fact, maternal stress can even reduce milk output, so that a nursing mother who is overstressed may be unable to meet the nutritional needs of her child.

The family unit forms a sort of collective environment that can be rich and supportive, or full of tension and dissension. Very little is known about the sensitivity of the newborn to family cues, but it is clear that even very young children are exceedingly sensitive to the nuances of family life. Family discord may lead directly to parental neglect, or it may cause the child to suffer indirectly, since an emotionally drained mother may be unable to provide adequate care for her child. If there is family discord, children may be drawn into it as a part of the power struggle between parents, or children may be excluded or marginalized, as neither parent may have the energy to meet the demands of their children. In either case, family discord can potentially have very damaging consequences for the children.

The larger social context around the family is a moderating influence that can be either benign or malignant, but is only rarely beneficial to a child. The social context can seldom compensate for an inadequate family life, but it can easily destroy the best efforts of a family to support and protect a child. Children born to the terror of the Warsaw ghetto or the Russian pogroms or the civil war in Bosnia may never reach their full potential because of the awful circumstances of their birth. On the other hand, there is evidence that children who are in a larger social context that is highly supportive, such as a kibbutz in Israel or a Head Start Program in Harlem, may derive little lasting benefit if their family life is already adequate. But, as young children grow into young adults, the larger social context becomes increasingly important. Teachers and friends assume an ever-larger role in the child's life, so that parents often begin to feel extraneous as a child reaches adolescence. The parental feeling of being extraneous is almost certainly inaccurate; parents remain important in the lives of their children even as these children reach adulthood. Nevertheless, it is almost completely unknown how and to what extent the family environment and the social context interact in fostering the maturation of a child, and how this interaction changes as the child ages.

In addition to these direct effects of the environment on the child, there can be a very complex interplay between genes and the environment. How a parent chooses to bring up a child forms the environment of that child, but it arises in part from the genome of the parent. Most parents provide their children with both genes and a home environment, and the two interact and reinforce each other in subtle ways. The interaction can be passive, such as when children receive both "smart genes" and an enriched environment from their parents. But there can also be an active interaction between genes and the environment; a child with a particular behavior can evoke something from the environment that actually reinforces the behavior. For example, children with high verbal ability are more likely to spend time reading, thereby further enriching their verbal ability. Or children who are more intelligent are more likely to test their environment continually, and to elicit more interesting responses from the people and things around them. Alternatively, children who are inherently somewhat hostile are more likely to evoke hostility from the people around them, so that hostility too can become self-reinforcing. And the interplay between genes and environment can take a lifetime to resolve; antisocial boys are more likely to experience social rejection as adults and have a higher than normal rate of divorce, unemployment, and criminal behavior.

How a child structures his own environment is critically important, and may radically affect how different children experience the same social context or even the same family. We are not passive participants in our own lives; we actively structure the environment around us all the time, and children are certainly capable of this as well. The structure a child chooses to impose on his environment is the structure with which that child feels most comfortable, and this structure is likely to have something of a genetic component.

Lead Exposure as a Paradigm for Environmental Influence

It is easy enough to claim that the environment can limit the success of a child in attaining its genetic potential, but what is the evidence for this claim? Is there any reason to think that the environment can actually constrain the expression of a child's full capacity? The easy (and politically correct) answer, of course, is "Yes," but the evidence is rather more fragmentary than one might expect. Perhaps the clearest example of an environmental effect on the expression of a genetic trait is seen with lead, a common environmental pollutant that can have a profound effect on human intelligence.

People have known for more than a century that lead is toxic at high concentrations; in fact, the Mad Hatter in *Alice in Wonderland* was supposedly driven mad by lead exposure suffered during the process of making hats. But it is only relatively recently that scientists discovered that lead, at levels often found in the environment, can also reduce IQ and can lower classroom performance. This conclusion is the result of studying more than 3000 children from Massachusetts, who were examined to determine the normal range of lead exposure, and the consequences of lead exposure at the upper bounds of the normal range.[19] A very clever research strategy was used for this; first- and second-graders were asked to give their shed baby teeth to their teacher, who then inspected each child to verify that there was a corresponding empty tooth socket. Shed teeth were chemically analyzed for lead content, with the idea that teeth are a stable and long-term record of lead exposure, whereas blood samples are likely to record transient or recent lead exposures. After the normal range of lead exposure was determined, children in the top 10% of tooth lead content were identified, and these children were compared to children in the bottom 10% of tooth lead content. Each child was tested for IQ and their academic performance and classroom behavior were rated by their teacher. In addition, scientists tried to measure every other feature of the environment that might conceivably affect IQ or interfere with school performance. High lead exposure was associated with reduced IQ, and specifically with reduced verbal ability and reduced ability to concentrate for long periods of time. Lead-exposed children were more likely to have behavioral problems at school, and high lead exposure was associated with distractability, impulsiveness, an inability to follow directions, hyperactivity, and poor overall cognitive function.

Over the years, this pioneering study has been confirmed by many other scientists who have been able to fill in some rather striking details. Lead exposure is more common among the poor, who often live in older buildings that have been painted in the past with lead-based paint. Since these buildings are less likely than normal to be well-maintained, or to have received a recent coat of paint, many inner-city homes are heavily contaminated with lead in the form of chips or dust. Contaminated soil is also a significant source of lead exposure in the inner city, presumably as a result of fumes from leaded gasoline. Children with high lead exposure are more likely to have a history of pica, a medical condition in which children eat inappropriate things such as paint chips or soil, which could, of course, account for their high lead exposure. Low-level lead exposure at birth is associated with reduced IQ at age 7, although environmental cleanup is associated with an increase in IQ among children who are moderately lead-poisoned. While current therapy to remove lead from the bloodstream seems to be relatively ineffective, time is an ally; environmental remediation can produce a long-term improvement in both blood lead levels and IQ. An increase in blood lead levels, from roughly the lowest third of normal to the highest 10% of normal, is associated with a reduction in IQ of about five IQ points.[20] Five IQ points may not seem like much, but millions of children are exposed to high levels of lead every year; this is perhaps tantamount to reducing the IQ of the entire country by a few IQ points on average.

While this research has been somewhat controversial, there is nevertheless very clear evidence that lead can reduce IQ and impair classroom performance. Since lead exposure has a grossly disproportionate impact on poor and minority children, lead poisoning may be responsible for some of the IQ discrepancies that have been reported in the past. Lead poisoning is thus a clear example of how an environmental effect may masquerade as a genetic effect. It is also an example of the often malignant effect of the environment on the individual; to envision an environment that has only positive effects on behavior is thus truly misguided. Toxic effects of the physical environment on the individual are legion, as mercury and several other heavy metals are also known to reduce IQ. But the truly frightening possibility is that many, perhaps even most, environmental effects on behavior have not yet been identified.

The Importance of the Social Environment

The environment is composed of both physical and social components, and we have seen that a toxic physical environment can limit the attainment of a child's full genetic potential. But it is much less clear whether or not a toxic social environment can also limit the attainment of human potential. Examples can be found in which profound social deprivation has a devastating effect on human development, but it is very difficult to determine whether moderate social deprivation is harmful. Similarly, it is virtually impossible to determine whether deliberate social enrichment is beneficial to the developing child. It often seems that everyone has simply assumed that social enrichment is beneficial, since profound social deprivation is clearly harmful.

There are several very colorful, perhaps apocryphal, stories of "wild children" who suffered the grim consequences of social deprivation. One of the earliest such stories, published in 1807, was of the "wild boy of Aveyron" who was captured in the forests of France at the age of about 12.[1] When captured, he could not speak or walk—he only shrieked and grunted, and crawled on all fours—and was totally indifferent to other humans. The boy was believed to have become retarded because of social isolation, so a kindly physician sought to educate him in human ways. After years of effort, the boy was able to discriminate among objects by sight or touch, to deal with simple abstract concepts, to write a few words, to respond positively to affection, and to respond to his new name "Victor." Although Victor was never able to achieve an intellectual level normal for his age, he did make great strides, and his achievements provided evidence for those who believed that mental development depends on a favorable social environment.

In 1942, J. A. L. Singh claimed in his book *Wolfchildren and Feral Man* that he had found two "wolf children" in India. Singh had been a missionary in 1920, when he heard tales of a "man-ghost" living in a nearby jungle. On inquiring about this apparition, he was taken to an abandoned termite mound, which was being used as a den by wolves. Singh had a hunters' blind set up near the den, and after several days he observed two female wolves with two wolf cubs and two human children, the latter completely covered with dirt and virtually unrecognizable. Several days later the children were caught, and both were found to be girls; one less than 2 and one about 8 years old. The younger child, named Amala by her captors, died within a year of being caught without ever showing significant human capabilities. The older child, named Kamala, lived for nearly 9 years in a missionary orphanage, and was able to undergo a partial transformation into a more human child. Initially, Kamala ran on all fours, lapped her food from a bowl, showed a preference for raw meat, and seemed more interested in farm animals than humans. She could not talk, but frequently howled like a wolf at night, and she was reported to be more vicious than a wolf cub. But Kamala gradually learned to walk, to use her hands, and to eat cooked food, and she also learned to show affection to some people. Eventually she learned a vocabulary of about 30 words before she died of a chronic illness. Her case suggests that an adequate social environment is required for proper development in childhood.

One of the strangest stories ever told about the consequences of social deprivation involves a young man named Kasper Hauser, who mysteriously appeared at the Nuremburg city gates in Germany in 1828. He was barely clothed, nearly mute, mentally confused, and strangely uncoordinated, although he seemed to have walked for many miles. No one could identify him, and he was unable to tell police anything about himself. He astonished the city fathers by trying to pick the flame off a candle, and further investigation showed that his mental faculties seemed to be intact but almost undeveloped. Over time, Hauser learned to speak and to write, and he eventually told authorities he had been raised in total darkness in a cellar, had been fed only black bread and water, and had never had human contact other than for a few moments each day when a man brought him food. His true identity was never discovered before he was apparently murdered 6 years later. As strange and perhaps apocryphal as this case is, it suggests the importance of the environment in fostering the development of a child.

There is also clear evidence that environmental deprivation in adulthood can be damaging. The Russians are alleged to have developed tortures based on sensory deprivation, using the principle that the absence of environmental stimuli causes sufficient mental anguish that physical pain is almost irrelevant. There is also evidence, from closer to home, that sensory deprivation can cause great mental distress. The solitary confinement or "boxcar" cells at the federal maximum security prison in Marion, Illinois, which are used for the most incorrigibly violent of prisoners, are small steel cubicles without natural light. The cells are insulated so that sound cannot get in or out, the ventilation is poor, and the only light is provided by a single 60-watt bulb. A prisoner who becomes ill has no way of alerting his captors, because the cells are so isolated from the rest of the prison. In fact, these cells are believed to be driving inmates insane, and several court cases have involved litigants who believe that incarceration in these cells constitutes cruel and unusual punishment.[7]

There is no doubt that an environment that falls within a normal and expected range of variation is important for development. As human beings, we are adapted at birth to respond to a certain range of possible environments, and as long as our environment falls within that range, the specific form of the environment may not matter very much.[21] Children are generally well-attuned to their environment, yet unaware of other possible environments, so that they cannot objectively assess the quality of their own environment; because of this, many children of poverty do not fully realize they were poor until they are older. In the absence of books and toys, family social interactions may become more important, so that environmental "complexity" is enhanced in other ways. Close proximity of neighbors in the inner-city may provide a rich stew of experiences that a child growing up in suburbia would never experience. While a city child may not often experience the quiet joys of a pastoral setting, he will experience a barrage of sights and sounds that would be unimaginable in the country. If a satisfying degree of environmental richness and complexity is available, it may not matter at all what particular form of stimuli are present. In other words, many different environments may be functionally equivalent to one another, as long as they fall within an expected range of variation. However, if an environment falls outside the normally acceptable range of variation, it might fail to promote proper growth and development of the child. This implies, of course, that many environments may be neutral, in terms of their effect on child development, while some few environments are actually detrimental.

Children are apparently able to structure their own environment, by picking and choosing among available stimuli.[21] Given a sufficiently varied environment, children are able to sort themselves into a personally satisfying milieu, according to their interests, their talents, and their personality. There is evidence that a child selects a particular environment because of heritable features of their personality, which implies that genes and environment are constantly interacting. The freedom to manipulate the environment increases with age, so there is an increasing ability to tailor the environment to fit the genes. The interaction between genes and environment continues for one's entire lifetime, and is likely to color career choice as well as choice of spouse and recreation. For example, few people would feel comfortable working in the maelstrom on the floor of the New York Stock Exchange, just as few people would enjoy working in isolation in a monastery. Many people might be uncomfortable in a confrontational occupation such as trial law, but many people would be equally uncomfortable in an occupation that permitted them no social contact. Similarly, few outgoing people would be interested in a relationship with an introverted, nonsocial person, just as few shy persons would enjoy a relationship with a highly extroverted "party animal." Each evening, most people have a choice of what leisure-time activity to engage in; the fact that some will elect to read, while others are watching television, using a Stair Master, or sleeping, is a result of genetics as well as environment. We are constantly presented with choices and the choices we make help to formulate our environment, so that every person experiences an environment that is based, at least in part, on their genes. All life experiences span a broad range, and the balance point we choose within that range may reveal much about our genes.

This idea has several rather startling implications. First, if most childhood environments are functionally equivalent, then efforts to enrich the environment may not be very productive. This further implies that while all parents may not be great parents, most are good enough. Ordinary parents are likely to have roughly the same impact on their children as those parents recognized as "superparents." Thus, beyond the contribution of "good genes," the parental role in creating "superchildren" is minimal; reading to an infant *in utero,* or playing Beethoven at the cribside, or trying to teach a child chess at age 5 is unlikely to have a lasting effect on development. This is a refreshing turnabout from the belief that parents have the power to ruin their children in so many ways. As Sandra Scarr, past president of the Society for Research in Child Development, says, "fortunately, evolution has not left development of the human species . . . at the easy mercy of variations in the environment. We are robust and able to adapt to wide-ranging circumstances. . . . If we were so vulnerable as to be led off the normal developmental track by slight variations in our parenting, we should not long have survived [as a species]."[21]

Another implication of the idea that children structure their own environment is that it may always be difficult to intervene successfully in the development of a particular child. In fact, if children actually do structure their own environment, they could actively resist a structure being imposed on them. Furthermore, intervention into a child's environment might not produce a positive result if his environment is already adequate; intervention could only help if the child's environment falls outside the range of "adequate." In other words, perfect nutrition will not make everyone an athlete, nor will a perfect school system make everyone an academic. The problem is that while it is reassuring to think that most environments are adequate, it is not known what actually constitutes "adequate." It may be that an adequate environment is common, or it may be that many environments are less than adequate; as of yet, there is no way to determine the adequacy of a particular environment.

In the 1960s, scientists studied adults who were tested as below average in infancy and preschool, but who had then received supplementary educational opportunities. They found that these people benefited greatly from the educational intervention, in terms of better grades in public school. This led scientists to conclude that a rich and varied environment is important for childhood development, and that the richer and more varied, the better. This was a major impetus for Project Head Start, an effort to enhance scholastic achievement of inner-city and minority children. Although the outcome of Head Start is generally assumed to be positive, several recent efforts to demonstrate this positive benefit have failed. In fact, it is not really known whether Head Start has been successful in the long term, or whether the degree of success has warranted the effort and expense. Nevertheless, programs like Head Start should not be abandoned, because it is possible that some children served by the program are suffering an otherwise inadequate environment, and these children might greatly benefit from the new opportunities. But we should not necessarily expect Project Head Start to increase the long-term scholastic attainment of all of its participants by a great deal.

5

New Tools for an Old Problem

Human behavior is the most complex and difficult problem ever attacked by geneticists, and it is likely to resist their best efforts for a full explanation for many years to come. In fact, it is quite likely that classical genetics would never be able to solve the most puzzling problems in behavioral genetics. This difficulty in understanding human behavior arises for several reasons. The first major problem for the behavioral geneticist is that human behavior is far more complicated than the simple physical traits that Mendel was able to study in his pea plants. When Mendel did his breeding experiments he could look at a pea seed and easily determine whether that seed was smooth or wrinkly, but human behavioral options cannot be so easily dichotomized. The second major problem is that, unlike most of the research for which classical genetics is best suited, human behavior does not involve one or a few genes. Instead, human behavior is determined by a large number of genes, each of which probably has a small effect, and there may even be many different combinations of genes that result in the same behavior. The last major problem for the behavioral geneticist is that the environment can have a profound effect on behavior, so that a genetic effect can be either accentuated or blunted by the environment.[22]

Given all of these major problems, it is a good thing that modern behavioral geneticists have at their disposal a range of new tools that could scarcely have been imagined even a decade ago. These new tools of molecular genetics will be described here, from the standpoint of both their strengths and their weaknesses. Only by having an adequate grasp of these tools are we able to determine the validity of what has been done with them.

The Brave New World of Molecular Genetics

The field of behavioral genetics has advanced with amazing speed recently because molecular biology has given scientists the ability to answer questions that could not have been asked a generation ago. It is now possible to prove that a gene causes a particular trait, even if one has no real knowledge of which protein is involved and no understanding of the particular function of that protein. The physical location of a gene on a chromosome can be determined with a fair degree of accuracy, even if there is no understanding whatever of what function the gene serves. This kind of genetic mapping is best for the simplest case, when a trait is determined by a single gene. But there is no reason in principal why genetic mapping cannot be used to dissect out the influence of several separate genes on a complex trait such as intelligence.

Success in modern behavioral genetics really requires three major areas of knowledge and skill:

1. Complete understanding of the basic patterns of inheritance. The various ways in which a trait can be passed down from parent to child must be understood, so that a mechanism of inheritance can be inferred simply by looking at a pattern of trait expression. This knowledge has largely been derived over the last century by studying the inheritance of simple physical traits in the fruit fly *Drosophila*. An understanding of the patterns of inheritance must include good insight into the complex ballet of the chromosomes during cell division. The way in which genes on a chromosome can rearrange and recombine themselves must be understood, because sometimes random rearrangements can cloud an

otherwise simple pattern of inheritance. Gene recombination is particularly common during the formation of egg and sperm cells, but it can also happen whenever a cell divides. Over time, recombination tends to break apart sets of genes that are linked, or adjacent to each other on a single chromosome. The extent to which genes recombine can be very informative in certain circumstances.

2. Facility with a range of methods to characterize the biochemistry of individual pieces of DNA. The modern behavioral geneticist must also be a biochemist, proficient with a variety of techniques that are used to characterize the biochemistry of DNA. These methods can be used to study the genetic makeup of an individual, or to find which gene codes for a trait, or to determine whether two people share the same version of a gene. Development of this branch of biochemistry has been exceedingly rapid over the last 20 years, which means that a behavioral geneticist must be adept at biochemistry.

3. Detailed knowledge of the pedigree of a large number of families. Thorough and intimate knowledge of the pattern of inheritance of a particular trait in a large number of different families is required. The trait analyzed in a pedigree could be IQ, or manic-depression, or the level of an enzyme in saliva. Information is usually needed on the general health and physical attributes of hundreds of related individuals in dozens of families. Pedigree information must go back at least three generations to be useful, because scientists are seeking to explain the appearance of a trait in a family pedigree based on their knowledge of the complex behavior of genes. Good pedigrees must also be detailed with respect to the bloodlines of a particular family. If a child is adopted (rather than born into a family) this must be known, since an adopted child can provide no genetic insight into his adoptive parents.

The workhorse of modern behavioral genetics is linkage analysis.[23] This approach is based on the assumption that related individuals sharing a trait in common also share a set of genes linked to that trait. The trait and the linked genes are both passed from generation to generation, and are almost always found together. These linked genes may (or may not) actually code for the trait in question, but they are at least physically close to that gene; thus whenever the trait is present, so are the linked genes. Even if linked genes do not code for a particular trait, their physical proximity to the relevant gene means that the trait and the linked genes almost always appear together. Because linked genes reveal the presence of another gene, they are considered to be markers for that gene. Scientists are most interested in finding a set of marker genes which have a known location on the chromosome and which are tightly linked to a particular trait. This will allow the chromosomal location of the trait to be determined, even if the gene for the trait remains unknown.

To explain this rather complicated concept by analogy, consider the association between tattoos and antisocial behavior. All tattooed people are not antisocial, nor are all antisocial people tattooed, but Hell's Angels often have large and garish tattoos. Therefore, the presence of an obvious tattoo can reveal antisocial tendencies, even if the person is not being obviously antisocial. Perhaps whatever personality traits induce a person to be antisocial also induce them to get tattoos, or perhaps another explanation is possible; the explanation is not critical. The key observation is that obvious tattoos are, in some sense, linked to antisocial behavior, and can serve as a marker for it. Unfortunately, marker genes are never as obvious as a large tattoo, but the principle remains the same: an obvious marker can reveal the presence of another less-obvious trait.

Linkage analysis simply proposes an explanation for the inheritance of a particular trait, then rigorously tests whether the proposed genetic model is also able to explain the inheritance of the marker genes. To increase the likelihood of success when doing linkage analysis, it is important to collect as many pedigrees as possible from families with a large number of members having the trait in question. Computer programs have been developed that can sort through a mountain of this sort of data very quickly, to help researchers focus in on the most plausible model of inheritance. Hundreds of human traits have been analyzed in this way, but the greatest successes have been achieved for traits coded for by a single gene.

Another approach to understanding the molecular genetics of a trait is a type of genetic dissection known as the allele-sharing method.[23] This method does not assume that individuals sharing a trait in common also share genes; it simply seeks to prove that two different individuals could not have gotten the same set of genes by accident. In other words, this method simply seeks to show that two individuals share genes more often than expected by chance alone. This technique has an advantage over linkage analysis in that it is less likely to be fooled; since it makes no assumptions about how genes were passed from one individual to the next, it cannot be easily wrong. Yet, allele sharing is also much less sensitive than is linkage analysis, so that it may fail to detect a relationship between a gene and a trait when such a relationship actually exists.

The genetic basis of a trait in humans can also be studied using a gene association method.[23] This method does not depend on family pedigrees at all, nor does it make any assumptions about a genetic mechanism of inheritance. It relies instead on identifying case individuals with a particular trait in common, and comparing these cases to control individuals who do not have the trait in question. The chromosomes of both cases and controls are carefully examined, to determine whether case individuals share a particular chromosomal pattern more frequently than do controls. Association methods have been used to compare those of high and low IQ, to determine whether there is a chromosomal pattern found at higher frequency in bright children. This type of study can be very powerful, as it is more sensitive than most other methods to rare traits. Yet this type of study is also vulnerable to error. For example, it can be shown that the use of chopsticks is associated with a particular type of protein on the surface of a blood cell. This does not mean that the blood protein causes a person to use chopsticks; instead, both the blood protein and the use of chopsticks are more commonplace in people from the Far East. Thus, to avoid falsely attributing a trait to a particular gene, an association study requires a large number of case subjects.

The Power of Molecular Genetics

Perhaps the best recent demonstration of the remarkable power of molecular genetics is a search through the entire human genome for those genes that make a person prone to diabetes.[24] Diabetes mellitus was first recognized in ancient Egypt, but it was not until this century that scientists determined that diabetes is associated with abnormal metabolism of the sugar glucose. There are two types of diabetes: insulin-dependent (Type 1) diabetes, which typically afflicts children, and non-insulin-dependent (Type 2) diabetes, which is common in older adults. Type 1 diabetes is associated with a near-total absence of insulin from the bloodstream, while Type 2 diabetes is associated with a loss of responsiveness to insulin. Insulin controls the ability of glucose to enter metabolizing cells, so when glucose is present but insulin is not, or when cells lose their ability to respond to insulin, the cells can literally starve to death in a sea of plenty. Type 1 diabetes is caused by the premature death of certain cells, located in the pancreas, whose normal role is to produce insulin. Apparently, these cells are killed when the body mounts an inappropriate immune attack against them, leaving the patient unable to secrete insulin. Because Type 1 diabetes runs in families and because it can afflict even very young children, it has long been recognized that there is a strong genetic component to the disease.

Linkage analysis has now shown that a whole host of genes are involved in susceptibility to Type 1 diabetes. Families were recruited for the study if at least two children in the same family had diabetes, and if the diagnosis was made at a young age; diagnosis had to occur before the age of 17 for one sibling, and before the age of 29 for the other sibling. Great care was taken with the selection of families to be studied, since scientists reasoned that many different combinations of genetic and environmental factors could potentially result in diabetes. By selecting study participants according to very rigorous criteria, scientists were able to ensure that they only studied families in which there was a strong genetic susceptibility to diabetes. A total of 96 families were identified in the United Kingdom, with each family having two diabetic children of the appropriate age, and each family having full kinship information available.

Analysis of these 96 families made it possible for scientists to focus in on chromosome 6 as a site where the most extreme differences from normal were clustered. A gene called *IDDM1* (an acronym for insulin-dependent diabetes mellitus gene 1), which is so strongly linked to diabetes that it was previously known, was found to be located on chromosome 6, in a part of the genome that controls immune recognition of self. But there were also other differences scattered across the genome, and scientists found that seven different chromosomes had at least one gene that was linked to diabetes susceptibility. To clear away this confusion, a total of 186 additional families, from both the United Kingdom and the United States, were added to the study. After analyzing the new families, it became clear that a total of 20 different genes may be involved in susceptibility to diabetes, with two genes being very, very strongly linked to the disease.[24]

These results are surprising for several reasons. While it is not at all surprising that two major disease genes are linked to diabetes, it is really quite surprising that as many as 18 other genes may be involved. At least 9 of the 20 genes are definitely linked to diabetes, with another 11 being more questionable.

Another surprise is that the 18 nonmajor genes could be identified at all, especially when using a fairly simple brute-force approach to gene identification. But perhaps the biggest surprise of this study is that there are no more than two major diabetes-related genes; the fact that so few major genes are involved in diabetes may make it feasible someday to attempt gene therapy for this disease. In addition, the fact that only two major genes are involved may mean that a genetic test for susceptibility to diabetes will soon be available. This would be useful because it might identify those persons who could most benefit from carefully controlling their diet. Finally, since the two major diabetes genes can be easily identified, it should now be easier to study the environmental factors that lead to diabetes.

The Problems with Molecular Genetics

When a tool or technique is brand-new, most people can easily appreciate the strengths of the new tool, but it is often considerably more difficult to see the weaknesses. For this reason, the limitations of the new tools of molecular biology are often not as widely recognized as are the strengths. This situation will, of course, change with time, as scientists confront the weakness of their techniques. Recently, several major behavioral genetics studies have been refuted (or at least weakened) because scientists were unable to repeat and confirm the results of earlier scientists. This has tended to cast a pall over the whole field of behavioral genetics, and some people have concluded that this infant branch of biology is premature or stillborn. Yet, while caution and conservatism are appropriate, recent developments in behavioral genetics should also engender real enthusiasm. Some mistakes have been made, as we shall see, but the problems that beset behavioral genetics are no worse, and really no different, from the problems that beset molecular genetics as a whole. Scientists will gradually learn how best to use the tools of molecular biology, to address fundamental questions of human behavior.

The problems that have plagued behavioral genetics are at their absolute worst in relation to schizophrenia and the mood disorders. These mental illnesses are subjectively defined, poorly understood, and often unsuccessfully treated. The diagnosis of schizophrenia is difficult because symptoms can be mimicked by a range of different conditions, including acute or chronic drug abuse, severe alcoholism, a brain tumor or other organic brain disease, and other mental illnesses. Consequently, a range of different conditions can be mistakenly diagnosed as schizophrenia. The diagnosis of manic-depression is perhaps worse; diagnostic criteria for the mood disorders keep changing, and each group of researchers tends to use a slightly different set of criteria to define who is affected and who is not. Both schizophrenia and manic-depression are complex illnesses whose symptoms are largely internal, so the symptoms must be reported to the physician by the affected individual. Many such individuals are not articulate enough to properly describe their condition, so diagnosis can be difficult. Furthermore, a range of different causes (either genetic or environmental) can all probably lead to the same consequence, so that even when schizophrenia is rigorously defined and properly diagnosed, there may be confusion as to the cause of the problem.

The problems in behavioral genetics, and in human genetics in general, are compounded by the fact that most people don't know very much about their own family. It may be generally known that Aunt Sophy died of cancer, but most family members won't know which of the several hundred types of cancer she had. Similarly, it may be well known that a certain relative had a debilitating mental illness or a personality disorder that made it difficult for him to get along with other members of the family. But even close relatives may not remember the exact diagnosis or the date of onset of symptoms. Furthermore, family bloodlines are often unclear, since illegitimacy was a source of great shame, and adoption by close relatives was commonplace, but not commonly discussed. The lack of effective birth control in prior generations, combined with the timelessness of adultery, sometimes meant that the birth father of a child was not the biological father. In many cases, immigrant families have completely lost touch with relatives in the old country, so that knowledge of kinship may extend back only one or two generations. Genetic analysis of subtle traits is far easier when there is extensive and multigenerational knowledge of a large number of families, but this is not usually possible.

The reason that large family kindreds are needed for behavioral genetics is that scientists are always concerned with showing that their findings are true and "significant." The degree to which a finding is significant is decided by using statistical tests to determine whether or not a finding could also have been obtained by random chance. For example, if a scientist proposes that a certain genetic marker is associated with schizophrenia, he is saying that the marker gene and the illness are found together more often than can be explained by chance alone. Yet the statistical tests used by scientists doing behavioral genetics are relatively new, and it is not always clear that they are being used properly. These statistical tests can be horrendously complicated and may require specialized computer software.[23]

The biggest statistical problem in behavioral genetics is that when you look at hundreds of gene markers in hundreds of people, sooner or later you will find a spurious relationship. Most people know that if you flip a fair coin, there is a 1 in 2 chance you will get a tails. The odds that you will get two tails in a row are simply ½ times ½, or 1 in 4. By the same logic, the odds of flipping 20 tails in a row are less than one in a million. But, if you flip a coin infinitely many times, you will certainly have at least one run of 20 tails. Similarly, if you study a huge number of gene markers, some of them will be found together with a trait, even if they do not actually cause that trait. This has been called a "repeated measures" problem, because the problem arises only if many measurements are made simultaneously. The only way around the repeated measures problem is to ask a very focused research question, so that it is unnecessary to make many measurements. Yet, to ask a focused question, the scientist must have a keen insight into the nature of the problem. This is often not possible; good scientists tend to ask questions for which there are no easy answers. In fact, if there was an easy answer, the scientist would probably not do the experiment at all. Research questions are interesting only if the answer is not already known. To a certain extent, if scientists really understand what they are doing, then it isn't science.

The Mathematical Analysis of Heritability

The best way to discern the separate effects of genetics and the environment on human behavior is to do a split twin study. As discussed earlier, this is the idea that identical twins, who share every gene in common, offer the best chance of determining the effect of the environment on expression of a trait. If identical twins are split at birth and reared in separate environments, then differences between these twins can arise only through the effect of the environment. Therefore, a measure of the similarity of split twins for a given trait is actually a measure of the power of genes to determine that trait. Of course, the researcher does not attempt to split identical twins in order to do this research. Instead, the researcher interested in this sort of question must search through records of adoption, in order to find twins who were split at birth. When information is compiled from hundreds of such separated twins, these data can be analyzed mathematically to determine the heritability of individual traits.

Heritability is defined as the amount of variation in a trait that can be explained by heredity alone. Simply put, it is a measure of the ability of genes to determine a trait; if the heritability is zero, then the trait cannot be inherited, whereas if heritability is 1.00, then 100% of the trait is determined by the genes. A heritability of 0.5 for a particular trait would thus mean that genes and the environment are equally important in determining that trait. In somewhat more mathematical terms, heritability is the proportion of "variance" in a trait that can be attributed solely to genetic variation. Heritability is calculated in such a way that variation related to the environment is factored out, and what remains is a measure of the impact of genetic variation on a trait. This calculation can be made in several ways, but each requires a great deal of information on the degree of similarity between members of a large group of people of known genetic relationship. The most direct way of calculating heritability for a particular trait is to measure the degree of similarity (i.e., mathematical correlation) between identical twins reared apart. This assumes that any similarity between split twins must have come from the genes they have in common; however, since identical twins share all genes, this is a reasonable assumption. This way of estimating heritability tends to produce a maximum estimate of heritability. Another somewhat more indirect way of calculating heritability for a given trait is to measure the correlation between identical twins reared together, and between fraternal twins reared together. Both sets of twins share environments, so the higher degree of similarity of identical twins reared together can only arise from shared genes. Heritability is approximated by calculating the correlation between identical and fraternal twins and subtracting one from the other; twice this difference is taken as a measure of the strength of the shared genes. This usually leads to a lower estimate of heritability than is obtained from the correlation between identical twins reared apart. The third method used to determine heritability is to use a mathematical model that incorporates all that is known about genetics to arrive at a more accurate estimate of heritability. While this latter method is still under development and is not firmly established as a method, modeling may eventually provide the most accurate estimates of heritability.

It is a critical point that heritability is a statistical abstraction from the data; thus, the concept of heritability applies to groups of people, not to individuals. While it is possible to state that personality traits are 41% heritable on average,[25] it is not possible to say that a certain person inherited 41% of his personality. In addition, the concept of heritability is inextricably linked to the conditions under which it is measured. If the environment changes, or if the genetic makeup of the population changes, then the estimate of heritability will also change. This can lead to

fairly large discrepancies between different studies in the estimate of heritability. Another problem is that many scientists
estimate the impact of the environment on the phenotype
without ever bothering to actually characterize the environment
in any way. This makes it impossible to determine what particular features of the environment are most important in determining phenotype. Furthermore, the mathematical approach used to
calculate heritability is open to several sources of experimental
error which can potentially cause major problems. Since heritability is usually calculated from split twin data, the calculation of
heritability will be incorrect if there are problems with the
design of the split twin study.

Split twin research has often been criticized because it is not
always clear that the twins were actually split at birth. In many
cases, twins were separated, but they continued to interact with
one another after the "split." This means that, to some extent,
the twins still shared a common environment. Any estimate of
heritability derived from such twins would tend to overestimate
heritability; because some of the environment is shared, these
twins will be more similar than otherwise. In addition to this
problem, many past studies of split twins involved relatively few
subjects, and the twins were often not representative of the
population at large. This bad luck or bad design could lead to
grossly incorrect estimates of heritability.

Given that there are potentially so many problems with
mathematical estimates of heritability, why is it useful or even
important to measure heritability? Many would argue that more
harm than good can come from new estimates of heritability, but
it is our viewpoint that heritability is still an important param-
eter to estimate. In the past, large-scale public policy decisions
have been based on crude estimates of the heritability of intelligence or mental illness. Recently, the authors of *The Bell Curve*
advocated many controversial changes in public policy, based on
their assumption that intelligence is 60% heritable. If this estimate of heritability is incorrect, then the public policy changes
advocated in *The Bell Curve* are grossly unfair. Thus, it is critically
important to obtain good measures of trait heritability, to ensure

that public policy is equitable and to undo any damage that has already been done. Estimates of trait heritability can only be improved if there is first improvement in the experimental design of split twin studies.

Modern versions of the split twin design tend to be more sophisticated than the old versions in several ways. Larger numbers of twins are typically recruited, and a greater effort is made to ensure that the twins were actually separated at birth. There is generally more effort made to ensure that subjects are representative of the population at large by accruing subjects randomly, so that a wider range of people with different racial and ethnic backgrounds are studied. Modern studies also tend to make a greater effort to follow the subjects over time, so as not to treat the individual like a static object. Many of the older studies did not characterize or measure the dynamic changes in an individual, which is analogous to studying a snapshot of someone in order to determine if that person is growing. Modern split twin studies tend to be longitudinal, meaning that many subjects are followed through time and reevaluated; this makes it possible to examine the effect of the environment on the development and expression of a trait. With all of these changes in study design, it is not surprising that modern estimates of trait heritability tend to be lower than some of the old studies. In any case, it is worth noting that even if the estimate of heritability of a trait is quite high, it is still possible to change that trait by changing the environment.

Meta-Analysis as a New Approach to Data Analysis

Meta-analysis is a catchy phrase for a mathematical tool that allows scientists to squeeze new information out of old data. While meta-analysis has not yet been used very much in behavioral genetics, it is inevitable that this powerful tool will soon find application in a field so plagued with conflicting studies. Meta-analysis is basically a way to combine the results of many separate studies, some of which may have been completed many years earlier, into one large new study.[26] Because a meta-analysis has many more subjects involved than any of the component studies, it is far more likely to be able to discern subtle trends and small effects. All that is necessary to do a meta-analysis is some statistical know-how and a research question with a long track record of published studies. These separate published studies are then reviewed, to determine the separate results and how best to combine them. When many separate studies all report a similar finding, even if they report a very small effect, then combining them into one large study can produce a strong finding. Meta-analysis is basically a clever way to deal with the most serious problem that confronts any scientist; namely, the problem of adequate sampling.

Most, if not all, scientific studies are too small to give a truly definitive answer to a question, simply because there is not enough time or money or manpower to do the "definitive study." But, by combining many separate studies, it is often possible to get closer to the ideal study. Recently, the power of meta-analysis was vividly demonstrated by a study that addressed a public policy issue related to behavioral genetics. In 1989, an economist at the University of Rochester reviewed the literature on educational reform, and concluded that there was no strong relationship between how much a school spent on educating its pupils and the level of success achieved by those pupils.[26] According to this review, smaller class sizes and higher teacher salaries did not translate into better educational outcomes for the students. Conservative critics hailed this work as showing that "money doesn't matter" in education. However, the study was badly flawed in that it weighted each prior study equally, gave each study one vote, and then simply tallied the votes. Thus, if each prior study had noted an identical small effect that was not statistically significant, then counting votes would still show no effect. A different approach was used in 1994, when a meta-analysis of the same 38 studies summarized in the earlier review came to very strikingly different conclusions. Using the more sensitive technique of

6

New Improvements in the Old Tools

The inheritance of behavior has been a subject of debate for many generations, but it is only very recently that the debate became grounded in observation and experiment. Prior to the use of the new tools of molecular biology, the debate about nature and nurture created much more heat than light. But the recent advances in molecular genetics, described briefly in the previous chapter, have now made it possible to address objectively questions that interest virtually everyone. In fact, some of the more pressing questions have even been given a satisfactory answer. This is not to say that any single question has been definitively answered; quite likely it will take another generation to really settle some issues. But scientists have made some real progress.

At the same time that molecular biology was blossoming as an approach to behavioral genetics, there were also some striking advances in the other tools used to study behavioral genetics. Some of these tools have been used for many years, but there have been cumulative improvements that now make the old tools new again. In fact, several of the old tools have improved so much that it is appropriate to think of them as essentially new tools.

Modern Psychological Testing Tools

The greatest weakness of psychology, and consequently also of behavioral genetics, has been the psychological testing tools used to measure human mental states, traits, and attributes. Scientists have been developing these psychometric tests for many years, and for many years have unwittingly incorporated their own preconceptions and biases into the tests. The first mental trait to be measured in a systematic way was intelligence, and this trait is still the most frequently measured; consequently, there is a richer tradition of mismeasuring intelligence than of any other human trait. During World War I, when new immigrants from Italy, Poland, and other parts of Eastern Europe were being inducted into the U.S. Army, the intelligence of these men was assessed by asking them to pick out the nicknames of professional baseball teams and to identify the product made by Smith & Wesson.[7] Those unfortunate immigrants who could not speak English were shown drawings and asked to identify the missing elements of the pictures. When a new immigrant was unable to point out that a tennis net was missing from a drawing of a tennis court, he was assumed to be intellectually inferior. This sort of absurdity persisted for many years, and left the whole field of intelligence testing open to quite justified charges of cultural bias.

Modern psychometric tests are the result of many years of effort by psychologists to strip away as much as possible of the cultural bias that was present in the early tests. Scientists have striven, with varying degrees of success, to characterize the aptitude for future achievement as well as to measure past achievement. This dichotomy is the basis for classifying tests as either aptitude tests or achievement tests. Aptitude tests try to measure abstract thinking and the ability to reason, aspects of "intelligence" that should have an impact on the ability to achieve in the future, even if past opportunities have been limited. Achievement tests try to take stock of the knowledge and skills already attained by the test-taker prior to the test. Obviously, this is a critical distinction; a gifted youngster growing up in deprived circumstances may have a great deal of aptitude, but not a great deal of achievement. The future potential of a child may or may not be realized, but it is crucial for our society not to limit the future of a child simply because the past of that child has already been limited. As a society, we are not so rich in talent that we can afford to waste any of it.

Current psychometric tests are more accurate than the old tests, in that they can more accurately predict later behavior or achievement. Current tests are also more likely to be suitable for persons from a range of cultural backgrounds within the United States. Yet most of the psychometric tests currently used in the United States assume that English is the language of greatest competency; a nonnative English speaker will usually test at a lower level than will a native speaker who is otherwise undistinguished. Although great progress has been made, psychometrics will probably always be a contentious field because human behavior is so ineffable, and consequently so difficult to characterize or to measure objectively. And human behavioral genetics will probably never be stronger than this weakest link; no matter how well something is correlated with "intelligence," this correlation is basically meaningless unless the measurement of intelligence is accurate. It is therefore appropriate to ask whether psychometric testing is necessary, or whether it actually does more harm than good.

The various arguments against psychometric testing have been clearly summarized recently[28]

1. Tests may have a cultural bias. This is by far the most reasonable and important objection to psychometric testing, and one of the most contentious questions in psychology today. The assumption has been made that psychometric tests are accurate for white, middle-class, Anglo-Saxon children, but that everyone else is given short shrift. Yet this assumption does not seem to be true, for the most part. Psychometric tests are tested for validity on a "normative population," which is a large group of children that includes racial and ethnic minorities in roughly the same proportion as those minorities are found in the population of the United States. This

ensures that, in the aggregate, the difficulty of test questions is about equal for everyone; some questions may be easier for white children than black, but other questions are easier for black children than white. When black and white children of the same socioeconomic stratum are compared, no significant differences are observed in intelligence. Basically, psychometric tests are not terribly good at predicting the future for anyone, but they are no worse at predicting the future of a black child than a white child.

2. Tests are not culture-free, even if they are not culturally biased. This is quite likely true; it is probably impossible to design a psychometric test that is entirely devoid of cultural content. However, items on a psychometric test represent important aspects of competence in the common culture. Failure to deal with these elements of the common culture on a test will almost certainly predict failure to deal with them in real life as well. For a democratic society to endure, all members of the society must be able to participate; this means that common cultural forms must be available to everyone. If we fail to ensure that everyone in the United States has equal access to the common culture, then what will result is a sort of cultural apartheid that cannot be fair.

3. National norms are not appropriate for everyone. It is true that someone from an ethnic minority should not be required to learn elements of a dominant culture that are unique to a different ethnicity. While the Battle of Culloden may still have resonance for someone of Scottish extraction, it is not appropriate that Polish-Americans be required to learn in detail about British perfidy to the Scots. Yet we all try to succeed in a common culture, and we should all be equipped for that struggle. It is entirely appropriate that schoolchildren in the United States learn about the American Revolution, even if their ancestors did not participate, because this is an element of the common culture that may explain

much about our society. As long as a test measures a skill or ability necessary for success in the common culture, minority interests are not well served by trying to eliminate national norms.

4. Minorities may be handicapped in test-taking skills. This is a real problem, because minorities may not recognize the importance of psychometric testing in the prevailing culture. If a test-taker is unable to read well, or to choose an appropriate strategy for problem-solving, or to balance the need for speed and accuracy, then the test score will likely reflect this. In addition, it should be remembered that psychometric tests are a social situation as well as a chance to excel. If a psychometric examiner is unable to make a test-taker feel at ease and welcome, the test-taker is unlikely to test well. This is more a reflection on the examiner than the examinee. Alternatively, if an examinee fails to recognize the importance of a test and treats it as a social situation only, this is also unlikely to produce good test scores.

5. Tests may deny access to adequate education. It is often true that psychometric test results are used to place someone in a special education class that offers no intellectual challenges, but test results are also used to advance someone into a gifted class. It is likely that mistakes have been made in placement, but similar mistakes would also be made in the absence of testing. An objective method of psychometric testing can eliminate favoritism and personal bias, and may also be more accurate.

6. Poverty interferes with the development of intelligence. Poverty, with its attendant risk factors of inadequate nutrition, poor health, deficient health care, fragmented families, wide availability of illegal drugs, childhood exposure to lead, alienation from the mainstream of society, lack of adequate support, low aspirations, profound feelings of worthlessness, and overt prejudice, is unlikely to be conducive to the development of intelligence. In

addition, children of poverty may be forced to assume adult responsibilites too soon, or a premium may be placed on their ability to learn skills unique to their cultural setting. Thus, poverty is unlikely to create a situation in which a test-taker can perform well. But the problem of poverty cannot be solved by eliminating psychometric tests; the best way to eliminate poverty is to improve educational access for all.

Counterbalancing arguments in favor of psychometric testing have also been summarized[28]:

1. Tests are useful to evaluate present function. A person cannot be helped without an accurate indication of what kind of help is needed. The diagnosis and treatment of problems and the determination of future prognosis is as critical to education as it is to medicine.
2. Tests can predict future function. Test scores can be very useful in allocating scarce educational resources, or in determining who would most benefit from an enrichment opportunity.
3. Tests are legally required to obtain access to certain special programs. This may or may not be wise, but it is the law. Until laws are amended, access to magnet schools or to special education classes is limited to those with appropriate test scores.
4. Tests are useful to evaluate the success of new educational programs. Without feedback from testing, it would be impossible to determine whether a new educational program is performing properly or serving a need. To abolish psychometric testing is to release the educational system from any form of accountability.
5. Tests can reveal inequality of educational access. Tests can provide an objective standard, so that claims of inadequate educational opportunity can be defended.

On balance, there seems to be a strong need to continue psychometric testing, at least within the context of the educational system. Whether or not psychometric testing actually serves a valid need outside the context of the educational system is a more problematic issue. Yet psychometric testing seems to be here to stay. Therefore, we need to assess carefully the validity of psychometric tests, to ensure that these tests are not culturally biased and to make the tests measure traits and behaviors more accurately.

A great deal of time and effort have been devoted to validating intelligence tests, because of the possibility that they are culturally biased and cannot accurately reflect the intelligence of ethnic minorities. Some very ingenious approaches have been used to determine whether cultural bias is real or imagined.[28] For example, scientists have compared the average test score achieved by different ethnic groups, after structuring the study group to ensure that all students tested were from the same socioeconomic stratum (SES). Overall, it has been found that intelligence of different ethnicities is not strikingly different as long as SES is held constant. Another approach is to determine whether intelligence tests are equally good predictors of achievement test scores for different ethnic groups. Generally, it has been found that intelligence tests are roughly equal predictors of success for all ethnicities, but that only about 30% of the variation in achievement test scores can be explained by differences in intelligence test scores. In other words, intelligence test scores, which cannot characterize the access to educational opportunity or the willingness to work hard once access is achieved, are equally poor predictors of achievement test scores for everyone.

Scientists have also examined intelligence tests to see if there is content bias; this would occur if questions on the test relied on prior knowledge unique to a particular culture. If a particular test item is disproportionately difficult for those of a particular ethnicity, this suggests content bias. As a rule, when test items are ranked from most to least difficult for each ethnicity separately, rankings tend to be the same. This implies that certain test items are more difficult for everyone, which is to be expected. Intelligence test items have even been examined by several judges, who were called on to rule on content bias. Afro-American, Hispanic-American, and Anglo-American judges were all equally unsuccessful at identifying biased test questions. One federal judge selected six test questions as being particularly difficult for black children, because he thought they contained items that were culturally biased. But when an equal number of black and white children were tested on these questions, the black children scored higher than the white children on three of the six items. This shows that even an educated and intelligent person cannot determine which test questions are biased. A great deal of evidence suggests that cultural bias on intelligence tests is minimal, and the fact that so many different studies have failed to find any objective evidence of cultural bias may mean that it is not present in current versions of psychometric tests.[28]

Another example of the validation that a psychometric test must go through is seen with the Scholastic Aptitude Test (SAT), used to predict which graduating high school seniors will do well in college. If the SAT is an accurate test, there should be a close correlation between the college performance predicted by the test, and the level of success actually attained in college. The SAT is validly used to predict college performance only if this correlation is close. For many years now, the predictive power of the SAT has been monitored, and research findings have been used as feedback to increase the accuracy of the test. Computer analysis of SAT results has been used to weed out those questions that do not distinguish well between high and low achievers. Such questions could fail to distinguish well between students either because everyone gets them wrong or because everyone gets them right. Other SAT questions have been eliminated because they do not separate good students from bad; if a question is typically answered incorrectly by someone who is otherwise doing well on the SAT, this is a bad question. Tens of millions of students have taken the SAT, and hundreds of millions of their answers have been analyzed, to winnow out questions that are invalid. The result is that the SAT test is now a reasonably good predictor of the achievable college grade-point average;

it is for this reason alone that the SAT is still required for most college applications. Similarly, modern psychometric tests have been revised and reworked for many years, to weed out invalid questions and to eliminate, as much as possible, cultural bias. The result is that there are now many different psychometric tests available, which are able to measure important personal variables with a reasonable degree of accuracy and predictive power.

Behavioral Insights from Laboratory Animals

Thirty years ago, long before it was clear that behavior can be genetically controlled, and at a time when it was more fashionable to think that all behavior is learned, a visionary scientist named Seymour Benzer, of the California Institute of Technology, began to do experiments with the fruit fly *Drosophila*. Benzer was interested in using a traditional approach to study a nontraditional problem; mutant fruit flies had been used for many years to study physical features, such as eye color or wing form, but no one had ever used them to study behavior. Benzer attempted to breed and study behavioral mutants, in order to gain insight into the inheritance of behavior.

The fruit fly is an attractive subject to use for this type of study, as noted earlier, because flies have a very short life span and mature very quickly. In addition, a breeding pair of flies can produce an astounding number of progeny in a short period of time, and the flies are relatively cheap and easy to maintain. For these reasons, the fruit fly has been the organism of choice for genetics experiments for many years. Benzer's experiments were crucial because they showed clearly that at least some aspects of behavior are genetic; yet fruit flies are so different from human beings that it is almost impossible to extrapolate lessons learned from flies to humans. More recent research with fruit flies continues to show that elements of fruit fly behavior are strongly influenced by the genes, but this really says very little about human behavior. For example, scientists have shown that a specific gene mutation, whimsically dubbed *fruitless*, controls the choice of who will be courted by a love-starved male fly. Males with the *fruitless* mutation court other males as enthusiastically as they court females, but then fail to complete copulation with any of the courted flies. A group of male *fruitless* mutants will form a bizarre chain of flies, almost like a conga line, in which each individual fly is both a courter and a courtee.[3]

Yet the fruit fly has also helped scientists to understand that sometimes "behavior mutants" are not what they appear to be. For example, a "behavioral" mutation of the fruit fly, known as *shaker*, causes a fly to tremble and interferes with the fly's ability to learn and remember. Clever research has shown that the *shaker* mutation is not a behavioral mutation; it simply makes the fly sick. Other fly behavior mutants are not ill, yet they still display fascinating deviations from the behavioral norm. For example, *period* mutants have an altered courtship song and they lose all semblance of day–night periodicity, yet they are otherwise healthy. The availablity of a wide range of fly behavior mutants has allowed scientists in some cases to actually study the biology of behavior. For example, one behavioral mutation has been attributed to an abnormality of a single protein in the fly nerve cell, such that the nervous system of the fly cannot function properly.

After Benzer pioneered the study of behavior with fruit flies, other scientists began to shift their attention to rodents with behavioral mutations. A clear advantage of this approach is that people are much more similar to rodents than to insects, so that it is easier to extrapolate from research on mice and rats to humans. Early experiments involved selective breeding for behavioral mutants, then determining which gene is associated with the mutant behavior. This meant that scientists were limited to accentuating natural rodent behavior, and could not create entirely new behaviors. More recently, it has become possible to study the genetics of behavioral genes at such a high level of sophistication that scientists literally use a mouse as a test tube. It is now possible to insert a single gene into a mouse to see what effect this has on behavior.[29] This is remarkable because most genetic crosses, even between two very closely related animals, involve hundreds or thousands of subtle genetic differences. By transferring a single gene into an invariant genetic background, gene transfer in a mouse can be extraordinarily informative. Because transfer is limited to a single gene, the function of that gene can often be unambiguously determined. For example, this technique was used to show that both long-term memory and maze learning in mice are dependent on a protein that acts as a receptor for another protein known as a neurotransmitter. Because the mutated protein cannot properly bind to the neurotransmitter, cell-to-cell communication in the nervous system of the mouse is impaired. Since all memory, whether it be maze learning in a mouse or musical memory in a human, depends on cell-to-cell communication among nerve cells, this mutation has a devastating effect on the memory of the mouse.

An alternative approach, that of selectively "knocking out" or deleting a single gene from the mouse genome, has also been used extensively in the last few years. In a sense, this is even more appealing than inserting a new gene into a mouse, since one cannot be sure that the newly inserted gene is properly regulated. In addition, if a gene is inserted into a genome, one cannot be sure that the function of preexisting genes is not somehow disrupted by the gene insertion. But if a single preexisting gene is selectively deleted, this is less likely to disrupt function of other genes. These "knock-out mice" have been used to study a wide range of different mouse behaviors, as well as several different diseases that are related to the mutation of a single gene. For example, it is now known that the circadian rhythm, or the day–night cycle of a mouse, is controlled by a single gene. When that gene is disrupted, the mouse loses any semblance of day–night periodicity and appears to waken or sleep randomly.

There is also a long history of studying primate behavior in the laboratory, in order to gain insight into human behavior.

Arguably, this research has never provided much insight into the role of genes and the environment in mediating primate behavior, since laboratory conditions are so different from the natural environment of the animal, yet some of the research has certainly been thought-provoking. Primates are intelligent enough that their behavior closely reflects human behavior, but they are also intelligent enough that their behavior is made aberrant by long-term cage confinement. Nevertheless, some primate experiments do seem to offer insights into human behavior. For example, scientists in the 1950s studied the way in which infant monkeys bond to their mothers. They were able to show that physical contact is required for proper emotional bonding with the mother, and that physical contact is more important than food in forming a strong pair-bond.

Laboratory research on behavioral mutants is very appealing because it allows the scientist to control many variables that cannot legitimately be controlled in a human study. However, there is always a great deal of controversy about whether or not findings from animal research can be applied to humans at all. Extrapolating from the behavior of a rat to that of a person is risky at best, and can lead to totally erroneous conclusions; extrapolating from a chimpanzee to a human is only somewhat safer. Of course, it will never be possible or desirable to do research on behavioral mutants in humans. But certain genes carried by rodents and humans are so similar that it is quite likely that the function of the gene is also similar. If proper care is taken to study a clearly focused problem, and to generalize from that focus only conservatively, the experimental analysis of behavioral genes in animals may have much yet to tell us about human behavior. A thorough understanding of the consequences of gene insertion or gene deletion in mice may eventually make it possible to treat, medically, certain genetically determined human diseases and eventually may make it possible to treat certain behavioral disorders as well. Yet this brave new world is far in the future; with a sufficiently long grace period before it becomes possible to treat behavioral disorders with gene therapy, we may learn to use this powerful technique wisely.

Sociobiology: Behavioral Insights from Wild Animals

Sociobiology is said to be the science of genetics in behavior, or the study of the degree to which evolution has acted on social interactions. It is a part of the trend away from laboratory studies of behavior, toward a more naturalistic and observational mode of study. The basic premise is that evolution acts to select social behaviors as surely as it acts to select physical features or mental abilities. The growth of sociobiology was very rapid in the past two decades, in part because of the publication of a textbook on sociobiology by E. O. Wilson of Harvard. Yet sociobiology, so popular a decade or two ago, has fallen somewhat out of favor, because of the tendency of certain scientists to make simpleminded pronouncements about human behavior based on their work with animals.

But sociobiology has led to a greatly heightened appreciation for the range and diversity of animal behaviors in the wild. There has been special emphasis placed on observational work with higher primates, because of their presumed similarity to humans. Many riveting films have been made about the field work of Jane Goodall with chimpanzees and Dian Fossey with gorillas. But the problem has always been that it is very difficult to generalize from animals to humans; it is not at all clear that the data on even the highest primates are truly informative about humans. And, of course, observing that a behavior is passed down from parent to child gives no insight at all into whether that behavior is genetically mediated or culturally mediated. This does not mean that sociobiology should be abandoned, because the field is a fascinating field in its own right, and may yet provide insights into human behavior. But it is also incumbent on us to be aware that we are different from even the higher primates. It is very difficult to generalize from the behavior of a mountain gorilla to that of a lowland forest gorilla; it is foolish to generalize from the behavior of a mountain gorilla to that of a human being.

New Applications for Old Tools

Several of the old tools that have been touched on here, including psychometric testing of intelligence and personality, and diagnostic testing for mental illness, have progressed greatly over the last several decades, and now seem to provide generally reliable measures of human traits. For this reason, psychometric testing will likely remain a part of the armamentarium of the behavioral scientist for many years to come. As it becomes possible to tease out ever-more-detailed information about the genetic basis of human behavior, precise and accurate methods to measure and assess this behavior will become even more valuable than they are presently.

But it is also likely that fewer and fewer direct connections will be made between human behavior and animal behavior, whether animal behavior is studied in the laboratory or in the field. Human behavior is so much more sophisticated than animal behavior that there can be little or no direct transference of insight from animals to man. Yet behavioral research with animals has already made two major contributions to our understanding of human behavior. First, research with fruit flies and mice has demonstrated unequivocally that even very subtle behaviors can be transmitted in the genes. This opens the door to speculation that human traits as intangible as altruism or artistic ability could conceivably be coded for in the genes. Second, sociobiology has shown clearly that evolution can act to select and modify the behavior of animals, and there is every reason to think that human behavior is also subject to natural selection. These two critical insights are reason enough to justify the continued study of animal behavior.

7

The Inheritance of Disease
A Paradigm for the Inheritance of Behavior?

We have learned a great deal about the heritability of various diseases in the last decade or two. Many of the major disease killers in the United States, including heart disease, cancer, stroke, diabetes, and atherosclerosis, are known to have strong genetic components, and the genetic basis of these diseases is now reasonably well understood. Perhaps the greatest increase in knowledge over the last decade has concerned the heritable nature of cancer.[6] We now know unequivocally that individuals in certain families have an elevated risk of cancer because they have inherited a genetic susceptibility to the disease. But the idea that individuals can also inherit a susceptibility to certain behaviors still seems heretical to many. Nevertheless, inheritance of disease seems to mirror the inheritance of behavior in many instructive ways; the lessons learned from studying disease inheritance even suggest several strategies for studying human behavior.

It is far easier to study the heritability of disease than to study the heritability of behavior, because disease is overt and usually easy to characterize. If a patient has cancer, that patient is generally quite ill, and the illness can usually be unambiguously diagnosed by a physician. Certain measurable features of the illness are present: the tumor can be visualized by magnetic resonance imaging (MRI) or by X-ray; the dimensions of the tumor can be measured; the blood chemistry can be analyzed; the presence of tumor antigens can be characterized; the growth rate of the tumor can be estimated; the genetic features of the tumor can be defined; or the clinical status of the patient can be described. This information can be used to compare patients, so that medical understanding of disease is continually improved and so that patient prognosis can be determined with accuracy.[30]

But, when dealing with behavioral traits, nothing is clear-cut. Many behaviors are difficult even for a psychiatrist to diagnose, and psychiatrists may differ in how they diagnose the same patient. Even the description of "normal behavior" is difficult, since there can be disagreement about what constitutes "normal." It is very difficult to measure behavior in an objective and meaningful way, and seldom is an easily measurable trait associated with behavior in the same way that easily measurable traits are often associated with disease. Our understanding of the mechanisms that produce behavior is thus very primitive relative to our understanding of the mechanisms that produce disease. Nevertheless, there is mounting evidence that much of human behavior is genetically determined. Therefore, it may be helpful to consider something of what is already known about the inheritance of human disease.

Studying Heritable Diseases in Groups of People

Epidemiology is the study of disease among large groups of people. Epidemiologists examine the spread of disease within communities of people, in order to identify the causes of disease. All too often, the cause of a particular disease is found to be familial. The heritable nature of cancer was recognized many years ago, but a great many other less-common diseases are also heritable. For example, Huntington's chorea is a devastating disease of the nervous system that causes progressive deterioration, dementia, and eventual death. Physicians have known for years that Huntington's is heritable and that the gene mutation that causes the disease is dominant. Thus, if one parent has the disease, on average, half his children will also have the disease. What is quite striking about Huntington's is that there is a tendency for disease expression to occur at a younger age in each succeeding generation. Scientists tried to find the genetic basis of Huntington's disease by studying a large family group from a town in the mountains of Venezuela; this family has been so badly ravaged by Huntington's that there was information about the disease going back several generations. In 1983, scientists used gene linkage analysis to show that the Huntington's disease gene resides on chromosome 4. Scientists then anticipated that it would be relatively easy to identify the gene causing Huntington's disease, because information was available from a family with a high incidence of disease, and because the linkage to chromosome 4 was identified rather quickly.

Yet this expectation proved to be completely wrong. It took six teams of researchers at ten different universities more than a decade of close collaboration to finally find the Huntington's gene. When the gene was finally found in 1993, scientists were still unable to understand how and why mutation of this one gene caused such a devastating disease. To date, the mechanism by which disease results from the mutation is poorly understood. The search for the Huntington's gene demonstrates very clearly that a gap exists between theory and practice; knowing the general whereabouts of a gene will not pinpoint the gene location. But research on the Huntington's gene reveals an even more frustrating aspect of molecular biology: knowing exactly where a gene is located does not reveal the gene function.

Cystic fibrosis is another heritable disease with devastating consequences, but it typically strikes young children, and often kills them before they reach adulthood. Symptoms of cystic fibrosis include acute respiratory distress, chronic vulnerability to infection, digestive tract problems, and malnutrition. The cystic fibrosis gene has also been localized to a specific chromosome, and scientists know that the gene codes for a single protein that is expressed in cells lining the passageways of the lungs. Scientists also know that this cell protein acts as a channel to let ions through the cell membrane. But, even with all of this knowledge, scientists still do not know exactly how a mutation of this gene causes the range of symptoms found in children with cystic fibrosis.

There is an instructive lesson here. Knowing which chromosome a gene is located on does not mean the gene will be easily pinpointed. Pinpointing where a gene is located does not mean that the protein coded for by the gene will be easily identified. Identifying the protein coded for by a particular gene does not mean that the function of the protein will be easily determined. And determining the function of the protein does not guarantee that it will be easy to understand the many and varied effects of changes in the protein.

In biology, form follows function. Form can even reveal aspects of function, but function can never be completely understood by simply studying form. In a sense, trying to determine the function of a gene using the standard tools of molecular biology is much like trying to determine the industrial output of Pittsburgh by looking at a road map. Without a doubt, a detailed map holds many clues about the function of factories in Pittsburgh; proximity to coal fields and to a major river suggests heavy industry, and street names would likely provide many clues. But a map, which is really a static representation of a place, is uninformative about the workings of that place. One can never know from a map how efficiently a place works, or how much work there is to do, or how one place interacts with other places.

The human organism is exceedingly complex and loath to give up secrets, and much of modern science is still descriptive. Molecular biology has indeed made great strides, but in many cases, the level of descriptive detail has increased without leading to substantial new insights. In other words, it is possible to know a great deal without knowing that which is most critical.

Genetic Susceptibility to Disease

A recent study of heart disease provides an interesting example of the role that genes play in human disease.[31] This study used a classic design in which twins are compared, to determine whether the occurrence of heart disease in one twin increases the risk of heart disase in the other. The Swedish Twin Registry was used, so that 21,004 identical and fraternal twins, born between 1886 and 1925, could be identified. Fortunately, this registry provides a fairly detailed life history and medical history for each twin, as well as providing the identity of the twins. The medical history of each twin was evaluated, as were the answers to a questionnaire that had been given in 1961 and 1963. A form of analysis, often used by insurance companies, was employed to determine which risk factors increased the risk of death from heart disease. Scientists wanted to determine whether the age at which one twin died of heart disease could be used to predict the age at which the other twin might also die of heart disease.

Several different risk factors for heart disease were evaluated, but age at death of one twin was generally the best predictor of death for the other twin. Conversely, factors such as age, body weight, medical history, or history of smoking were not able to predict age at death from heart disease with much accuracy. Generally, the concordance for heart disease was higher for identical twins than for fraternal twins, as one would expect if heart disease were strongly genetic. For example, if one member of a set of male identical twins died before the age of 55, the risk that the other twin would also die before age 55 was increased more than eightfold with respect to normal. Yet if one of a set of male fraternal twins died before the age of 55, the risk that the other twin would also die before age 55 was increased only fourfold. In other words, identical twins are more than twice as likely as fraternal twins to be concordant for age at death. Strangely, female twins are more likely than male twins to be concordant for death from heart disease, perhaps because it is relatively rare that a woman die of heart disease. If a woman dies of heart disease before age 55, her identical twin is about 15-fold more likely than normal to die of heart disease. The genetic effect on heart disease generally decreases with age; the older a person is when they die of heart disease, the less likely it is that heart disease was hereditary.

We expect that most diseases will have both a genetic and an environmental cause, but that the two components will be essentially impossible to separate in the individual patient. This is why the epidemiologist studies large groups of people. It is only by looking at groups of people, and by somehow controlling for genetic influence, that scientists can determine what proportion of disease incidence is caused by the genes and what proportion is caused by the environment.

One possible way to control for genetic influences is to compare the cause of death in adopted children to the cause of death of the biological parents.[32] Adopted-away children do not share any environment with their biological parents, so this should eliminate all environmental similarities between biological parent and child. This type of study has been done using the Danish Adoption Register, which contains records of all adoptions formally granted in Denmark during the period from 1924 through 1947. A group of 960 adoptees was culled from the registry, and the medical history of these adoptees was compared to the medical history of both the biological parents and the adoptive parents.

The cause of death for adopted children is often more similar to that of their biological parents than their adoptive parents.[32] This strongly suggests a genetic cause for death, especially in cases where death occurs prematurely. An adoptee whose biological parent died before the age of 50 is almost twice as likely to die prematurely from "natural causes." If infection alone is considered, premature death of the biological parent increases the risk of death of the adoptee almost sixfold. Of course, this does not mean that in the cases studied, infection was inherited; in most cases, it was many years between death of the parent and death of the child. Rather, this means that somehow the adopted child inherited a susceptibility to fatal infection. In addition, if a biological parent died prematurely of heart disease, the risk of premature death for the adoptee was more than fourfold higher than normal. This means that, while heart disease is known to be related to dietary factors, the death of a parent from heart disease predisposed a child to death from heart disease no matter what the diet. On the other hand, premature death of an adoptive parent did not significantly increase the risk of death for an adoptee, unless the adoptive parent died of vascular disease or cancer. If an adoptive parent died prematurely of cancer, the adopted child was roughly fivefold more likely to die prematurely of cancer. This finding strongly suggests that cancer and heart disease are both caused by exposure to environmental risk factors.

One could dismiss these findings by saying that science has merely proven that death runs in families, but this would be trivializing several very important studies. Instead, the various trends suggest a complex interplay between genes and the environment. Some genes may be so strongly predisposing to a particular disease that the environment probably does not much matter. Other genes are probably only weakly predisposing, so that there must be some environmental factor that participates in producing the disease. Still other diseases seem to be unrelated to genetics, and instead are solely related to some feature of the environment. It is only by examining disease incidence in twins, or by using some other sophisticated approach to control for genetic influence, that we can discern the relative role of genes and the environment in producing disease.

The Genetics of Disease in Split Twins

The best indicator that a disease is strongly hereditary is if identical twins are both prone to the same disease, even though fraternal twins are not. The distinction between identical and fraternal twins is very important; if all twins were subject to the same disease, this would imply that the disease is environmental. All twins share environments, so a disease that is environmentally mediated should be common to all twins. On the other hand, only identical twins share all of their genes, so only identical twins are expected to be highly concordant for a disease that is genetically mediated. This is not meant to imply that fraternal twins do not get the same genes; since fraternal twins have the same parents, they share at least some of the same genes. But identical twins share 100% of their genes, so they should share 100% of the genetic problems that come with their genes.

The strength of the genetic contribution can be crudely estimated for various diseases, using an approach that is perhaps just a bit too simplistic.[2] If the similarity between identical twins is known for a particular trait, this concordance must be the result of identical genes and a highly similar environment. But concordance between fraternal twins for a particular trait occurs with fewer shared genes, so it is more a measure of shared environment. To approximate the genetic effect, the concordance for fraternal twins is subtracted from the concordance for identical twins, then the result is doubled. For example, the concordance for a certain type of epilepsy (known as idiopathic epilepsy) is about 83% in identical twins, while the concordance in fraternal twins is about 30%. Heritability for this type of epilepsy is therefore roughly 100%, or twice the result of 83 minus 30. This, of course, does not mean that 100% of people with epilepsy genes will get the disease; we know this is untrue since only 83% of identical twins are concordant for epilepsy. A better way to think of it may be that 100% of people with the "epilepsy gene" are at risk, but some of these people are in some way protected from disease. Thus, while epilepsy is strongly heritable, the environment can still prevent the epilepsy gene from being expressed.

Clearly, this method of calculating heritability is flawed, since it would imply that 106% of idiopathic epilepsy is heritable. But the general principle is that comparing disease incidence in identical and fraternal twins can give a good idea of the strength of a genetic influence. The exact value of heritability is not as important as the relative value, compared to that calculated for other diseases. Using this method to estimate heritability, rheumatoid arthritis is also found to be very strongly familial, since 50% of identical twins are concordant for arthritis, but only 12% of fraternal twins are concordant.[2] By the same method of calculation, the heritability of peptic ulcer and hypertension is lower, but still substantial. Heritability of heart disease is lower still, while that of breast cancer and Parkinson's disease is very low indeed.

Overall, we have seen strong indications that human disease is hereditary, to at least some degree. In the next several chapters we will consider evidence that certain human behaviors, including intelligence, retardation, schizophrenia, autism, manic-depression, reading disability, attention-deficit disorder, and even homosexuality, can also be hereditary. But if human behavior is truly hereditary, it would seem likely that we can learn something about the inheritance of behavior by studying the inheritance of disease. In other words, what analogies can be made between inheritance of disease and inheritance of behavior?

1. Behaviors Tend to Aggregate in Families

Clear evidence has been obtained that there are "cancer families" prone to a wide range of cancers. For example, Li-Fraumeni syndrome causes an increased susceptibility to cancers of the muscle, breast, brain, and bone in children and young adults. The syndrome was discovered by an astute physician who noted a cluster of rare cancers in a single family. In this family, three young patients developed a type of muscle tumor that ordinarily afflicts only one in 100,000 persons in the general population. On questioning the parents of the afflicted children, it was found that two of the mothers had breast cancer at an early age, one father had leukemia, and other family members had developed a range of cancers at an early age. No environmental factors seemed to account for all of these cancers, so it was recognized that these patients were members of a "cancer family."[6]

Similarly, there is strong indication that manic-depression, one of the most common of all mental illnesses, also aggregates in families. Manic-depression can cause debilitating mood swings, with the sufferer alternating between a state of crushing depression with intrusive thoughts of suicide, and a state of euphoria characterized by hyperactivity, irritability, and irresponsibility.

Alfred Lord Tennyson, the British poet, was prone to long periods of recurrent depression, and he came from a family so heavily afflicted with manic-depression that he himself recognized the illness as a "taint of the blood."[33] Tennyson's father, his grandfather, two of his great-grandfathers, and at least 8 of his 11 siblings were similarly afflicted with symptoms of manic-depression. Tennyson's brother Edward was confined to a mental asylum for 60 years before dying from exhaustion after a fit of mania. The entire family was prone to episodic depression, irritability, and eccentricity, and even the siblings who did not have full-blown manic-depression were still not entirely normal.

2. "Genetic Background" Can Modulate Trait Expression

Scientists are beginning to suspect that inheritance of simple human traits may not be so simple after all. Each of us has a unique genetic background against which a trait is expressed; the vast collection of genes we are born with can apparently interact with one another in remarkably subtle ways to modify the expression of a particular trait. Perhaps the clearest example of this is seen in the inheritance of sickle-cell disease, which is caused by a mutation in the gene that codes for hemoglobin. People with this mutation have abnormal red blood cells that are unable to carry oxygen as well as can normal blood cells, so they often show signs of a chronic oxygen deficiency in their tissues. Some patients suffer devastating consequences from this mutation; children only a few years old can suffer a massive stroke that paralyzes or kills them, and most patients endure repeated painful and debilitating sickling crises every year. Yet some patients with sickle-cell disease are very nearly symptom-free. In fact, some years ago, a massive screening effort in Brooklyn identified two patients with sickle-cell disease who had never been ill from the disease and who didn't even know they had it. Evidently, something in the genetic background of these two patients prevented them from suffering symptoms, even though they were afflicted with sickle-cell disease.

Furthermore, even among symptomatic sickle-cell patients, the severity of the symptoms can vary widely. Patients who have ostensibly identical mutations can suffer very different symptoms. Some patients have a stroke at a very early age, while other patients never suffer a stroke. Some patients have symptoms that are predominantly pulmonary, while other patients suffer recurrent infections instead. Some patients require frequent hospital stays, while others only rarely need to see a physician. The only possible reason for these variations in symptom expression is that there is something in the genetic background of some patients that modulates the severity of symptoms of the disease. The fact that genetic background can so strongly affect the expression of a disease implies that genetic background can also affect the expression of a behavior or trait. For example, there is some indication that genetic background can have a strong impact on the age at which Alzheimer's dementia is diagnosed in patients.[34]

3. The Younger the Age at Diagnosis of a Trait, the Greater the Role of Heredity

Certain cancers are strongly familial, and the younger a patient is when diagnosed, the more strongly is genetics implicated.[6] In fact, there is clear evidence that some children are actually born with cancer, while other children are prone to a wide spectrum of familial cancers. There are now known to be at least 50 different hereditary cancer syndromes, in which there is an often dramatic familial increase in cancer incidence. Children with Li-Fraumeni syndrome have a greater cancer risk than most adults, and these children are 21-fold more likely than other children to get cancer. It is almost as if these patients were doomed to have cancer from the start, with the greatest uncertainty being the age at which cancer will be diagnosed.

Another well-known example of an inherited cancer is retinoblastoma, a cancer of the eye that afflicts young children.[6] This cancer is associated with a mutation of the retinoblastoma gene, and the mutation is passed from parent to child. A child born with the retinoblastoma gene mutation has a 95% probability of developing retinoblastoma, whereas a child without the mutation has only a 1 in 30,000 chance of developing this cancer. Thus, retinoblastoma risk is 28,500-fold higher in children with the mutated gene.

Autism, a condition that can cause extreme social withdrawal in children, is also thought to be very strongly heritable. It typically afflicts very young children as well, so young, in fact, that there has arguably not been time for the environment to have an effect on the child.

4. The Older the Age at Diagnosis of a Trait, the Greater the Role of Environment

Cancer in older patients is frequently a disease of abuse or disuse, and up to 60% of cancers that affect adults could be prevented by limiting environmental exposure to carcinogens.[6] As a general rule, the older a patient is when diagnosed with cancer, the more likely it is that the cancer was caused by something in the environment. The converse of this rule is that the older a patient is when diagnosed with cancer, the less likely it is that the cancer was inherited. Analogy suggests that traits that are expressed solely in older adults are likely to have a larger component of environmental influence. In fact, there is evidence that Alzheimer's dementia has an environmental component, since identical twins are not always concordant for this condition.

5. Behavior Results from an Interaction between Genes and the Environment

We have seen that inheriting a retinoblastoma gene mutation increases the risk of this cancer 28,500-fold with respect to normal. Yet about 5% of patients who carry this mutation never develop the disease. Furthermore, retinoblastoma patients typically develop no more than three or four tumors among the millions of cells in the retina, even though all retinal cells carry the mutation. Therefore, even for this strongly hereditary disease, cancer formation appears to be a relatively rare event that requires some sort of environmental interaction.

Similarly, a genetic analysis of 337 families affected by lung cancer has confirmed that there is a genetic predisposition to lung cancer.[35] However, the best predictor of lung cancer risk is the total number of cigarettes smoked, and lung cancer is rarely diagnosed in the absence of exposure to tobacco. Thus, the genetic predisposition to lung cancer is inherited, but the trait is apparently expressed only when the gene interacts with exposure to tobacco smoke. Nevertheless smokers with no family history of lung cancer cannot conclude that they are resistant to the cancer unless both of their parents also indulged in long-term tobacco use without evidence of disease.

By analogy, we anticipate that most behaviors will be expressed only when environmental conditions permit or enable expression. For example, intelligence is strongly heritable, but it can only be fully expressed in a child who experiences early sensory stimulation, proper medical care, adequate nutrition, and good schooling.

6. Susceptibility to a Behavior Is Inherited, Not the Behavior Itself

Scientists know that familial cancers are usually related to an inherited susceptibility, not to an inherited disease.[6] Those cancers that are directly inherited are quite rare (e.g., retinoblastoma), while cancer susceptibility has been linked to some of the most common cancers, including colon cancer and skin cancer. Similarly, behaviors are probably the result of an inherited predilection, not of an inherited behavior. For example, a proclivity to violent behavior might be associated with low intelligence, good physical health, irritability, and a low tolerance to frustration. Thus, violence itself is almost certainly not directly heritable, although several conditions that might predispose to violence are heritable. In other words, the inherited component of a behavior may take the form of a built-in bias in the response of the individual to the environment. But individual expression of a behavior is usually so variable that it is difficult to imagine how the behavior could be inherited without invoking a role for environment. And if the environment does play a substantial role in behavior, then the inherited component alone must be insufficient to cause behavior.

7. The Difference That an Individual Gene Makes Is Usually Rather Small

Earlier, we saw that diabetes is associated with a change in as many as 18 different genes, with only two of those genes having a major effect. A mutation in one of the 16 non-major genes is thus likely to have a rather small effect on the onset or severity of diabetes. Similarly, most behavioral genes probably increase the likelihood of a particular behavior by a relatively small amount. This will be especially true if several or many genes are involved in the expression of a particular behavioral trait. For example, scientists have guessed that at least five to seven genes are involved in determining intelligence. If this is true, then a change in one of these genes would be expected to result in a rather small incremental change in intelligence quotient (IQ).

8. The Best Insight into Heredity Will Be Obtained for the Rarest Behaviors

The first cancers to be identified as hereditary and traced back to the gene level were all quite rare. This, of course, does not mean that all cancers are rare, or that cancer is caused by a rare gene. It simply means that it is easier to find a gene that causes a highly unusual cancer. This is because the person with a highly unusual cancer is likely to have a highly unusual gene mutation.

Similarly, if a gene causes a very common behavior, it will be hard to identify, simply because too many people will have the gene. Scientists are much more likely to identify a gene if it codes for a very unusual behavior or ability. For this reason, researchers interested in the genetics of intelligence have begun to focus on children of very high or very low intelligence.[22] By comparing highly intelligent children to children of low intelligence, one is comparing extremes, so that it is much easier to identify those rare genes associated with great intelligence.

9. A Single Gene May Predispose an Individual to Many Behaviors

People with Li-Fraumeni syndrome are vulnerable to a wide range of cancers, with a risk of pediatric cancer 21-fold higher than normal and a lifetime risk of cancer about twofold higher than normal.[6] Although cancer risk is most profoundly elevated for a few cancer types, there is also a higher susceptibility to a broad range of cancers. Among patients with Li-Fraumeni syndrome who are less than age 45, at least 87% of all cancers can be blamed on this syndrome.

By analogy, one would expect that a single gene might also predispose to a wide range of seemingly unrelated behaviors. In fact, there may already be evidence for this idea, since it was reported in 1991 that one gene was significantly associated with a wide range of neuropsychiatric disorders.[36] This gene, called the A1 allele, was found at a significantly higher incidence than normal in people suffering from autism, attention deficit–hyperactivity disorder, Tourette's syndrome, and alcoholism. In addition, there may have been an association with posttraumatic stress disorder, schizophrenia, and drug addiction, although these associations were not statistically significant. The A1 allele was thought to work primarily by modifying the effect of another gene, rather than by directly causing all of these conditions, but the range of associated disorders is stunningly broad. However, this was a very controversial study, so it is possible that these associations may be disproven in future studies.

10. It Is Critically Important to Understand the Inheritance of Behavior

A better understanding of the inheritance of cancer will help us to determine which persons are most at risk of getting cancer.[6] It should be possible, within the next decade, to predict accurately an individual's risk of developing cancer in the future. This individualized prediction of risk is vitally important, because it will enable physicians to individualize cancer prevention strategies so as to do the most good. For example, a woman known to be at elevated risk of breast cancer would be advised to undergo frequent mammographic examination, whereas a man at higher risk of prostate cancer would be advised to get a prostate-specific antigen (PSA) test more frequently than normal.

Similarly, if the inheritance of behavior was better understood, this might enable scientists and policymakers to focus attention where it is most needed. It might become possible to design behavioral interventions that actually work, instead of the current situation, in which countless interventions have been politically mandated, only to fail. An intervention that truly benefits the individual will also benefit society, whereas an intervention that fails can potentially harm both the individual and society.

11. It May Become Possible to "Screen" for Behavior

It is very likely that a number of new cancer screening tests will be introduced or used more widely over the next few years.[6] These screening tests will be able to identify someone at risk of cancer before the cancer becomes established. An excellent example is the PSA test, which has already succeeded in identifying many early-stage prostate cancers. In general, when a cancer is identified early by a screening test, the likelihood of cure is much higher. The best of the new cancer screening tests will probably save many lives.

By analogy, as genes that predispose to behavior are identified, it should become possible to identify those persons most at risk of certain behaviors. If it became possible to screen for persons who are manic-depressive, this could potentially save lives because it would enable physicians to intervene medically before the condition became life-threatening. Roughly 20% of untreated manic-depressive patients commit suicide, and between 60 and 80% of all adolescents and adults who commit suicide have a medical history consistent with manic-depression.[33] If these people received treatment for their condition, this extraordinarily high incidence of suicide would almost certainly decline.

12. Race May Be a Risk Factor for Certain Behaviors

There has been an increased emphasis on studying disease in diverse ethnic groups, and new evidence has been obtained that cancer risk differs among the races.[6] These differences may mean that race is actually a risk factor for cancer, especially in the presence of other risk factors. This type of study is very controversial, and very difficult to do properly, because Americans are so sensitive to race as an issue. Nevertheless, these studies are important, precisely because race should not be an issue in determining health care delivery. If all races are to have equal access to quality health care, then there must be a better understanding of cancer risk factors for all persons, regardless of race or ethnicity.

Similarly, it is possible that racial differences could put people at a differing risk for certain behaviors. Trying to identify an association between race and behavior is likely to be highly contentious, but this does not make it morally wrong. It is not racist to recognize that there are differences between the races, although this recognition can be used for racist purposes. It may be somewhat naive to hope that knowledge can be used for the benefit of all races, but it is very clear that the lack of knowledge can be used to oppress a race.

13. Some Genes May Act as Behavioral Suppressors

Patients with Li-Fraumeni syndrome are potentially vulnerable to a wide range of cancers. This is because the normal role of the Li-Fraumeni, or p53, gene is to suppress cancer. The p53 gene has been called the guardian of the genome, since its normal function is to suppress growth of aberrant cells that could form tumors. When p53 is mutated, it loses this ability to suppress cancer, so that other cancer-causing genes can be expressed.

It is impossible, at this point, to determine whether there are also behavioral suppressor genes. But it is possible to conceive of a role for a behavioral suppressor gene. Perhaps the mood swings of a normal person are dulled by the presence of a suppressor gene that blunts the emotional response to life events. If this "emotional blunter" gene became mutated, the wild mood swings of manic-depression might result. While this possibility is highly speculative now, it is an interesting idea.

14. Spontaneous Mutation May Give Rise to Behavior

A great deal of research has demonstrated the importance of spontaneous mutations in creating cancers. Analogy suggests that spontaneous mutation may also be important in creating behavior. How else can one explain anomalous individuals such as Jeffrey Dahmer or John Wayne Gacy, who are so far from normal that normal people cannot begin to understand them? The problem is that research into spontaneous mutations of behavior is virtually impossible to do in human beings.

Our current approach to separating the influence of genes and the environment would not be sensitive to the possibility of spontaneous behavioral mutations. In fact, it is quite likely that a spontaneous behavioral mutation would be misattributed to an environmental effect, since most scientists have not yet considered the possibility of behavioral mutation. Perhaps what we now construe as "unshared environment," the environmental differences experienced by siblings within a family, is contaminated to some extent by spontaneous mutations. It is perhaps possible that the real genetic component is actually comprised of the now-known genetic component plus part of that component currently ascribed to unshared environment. This idea, of course, is even more wildly speculative than the idea of a behavioral suppressor gene.

These speculative but intriguing parallels between disease and behavior cannot yet be verified, because our knowledge of human behavior is just too rudimentary. Yet making analogies between disease and behavior seems legitimate, since both conditions are known to have genetic as well as environmental causes. Our knowledge of human disease so greatly exceeds our knowledge of human behavior that the lessons learned from studying disease may well serve to guide and illuminate our studies of human behavior.

8

Intelligence

The inheritance of intelligence, and the policy implications that this may or may not have, is currently the most controversial and divisive issue at the interface between science and society. This is in part because of *The Bell Curve*,[37] which has given respectability to the notion that it is right and proper to discriminate against the disadvantaged. While the book is badly flawed, the authors should perhaps be applauded for catalyzing many scientists more qualified than themselves to address publicly the inheritance of intelligence. But the book should also be seen for what it is: a political agenda masquerading as science, a mean-spirited diatribe against the poor and disenfranchised, and a pseudointellectual legitimization of racism. Racism is revealed, not in recognizing that racial differences exist, but in judging that some racial traits are better than others, and in believing that all racial traits are genetically fixed and immutable.

The reason why the inheritance of intelligence is such a volatile issue is clear. But the reason why so little has been said about intelligence by reputable scientists in the past decade is not clear. Scientists are often reluctant to leave the cloistered environment of the laboratory or the lecture hall to confront the issues of the day. This reluctance may arise because scientists are not interested, or because they feel unqualified, or because they are shy, or simply because they think the public is not ready for science. Most scientists are accustomed to seeing stories about science reported in the lay press that are so badly mangled or oversimplified or out-of-context that they are no longer true. Yet, by their reluctance to confront the issues of the day, scientists have allowed themselves to be exploited by the writers of *The Bell Curve*.

It has also been intellectually fashionable, for many years now, to emphasize the importance of the environment in determining human behavior. In fact, many scientists have mistakenly regarded humans as a *tabula rasa*, a blank slate on which experience writes. In this view, the genetics of human behavior is of no importance, since all behaviors are learned. This viewpoint first gained respectability after World War II, when scientists became aware of the unspeakable horrors of the eugenic program practiced by the Nazis. The environmentalist viewpoint was firmly entrenched during the 1960s, when it seemed that all things were possible for Americans; we could triumph over the Nazis and eventually the Russians, we could land on the moon, and we could certainly triumph over our genes. Psychology was enamored with the idea that good behavior could be programmed into the individual, and psychiatry was fixated on individual experience as the key to all mental illness. In this context, it is no surprise that the genetics of behavior was downplayed.

The first real crack in the armor of environmentalism was sociobiology, a set of ideas that were introduced to the public in the mid-1970s. Sociobiology claimed that animal behavior is strongly hereditary, and that we can gain insight into human society by studying animal societies. These ideas were a rude slap to many, because of the long-standing and very comfortable belief that we, as humans, can rise above our genes. One of the most vigorous defenses of environmentalism can be found in the book *Not in Our Genes*, which was cowritten by Richard Lewontin, a very prominent geneticist. This book is a well-researched and well-presented attempt to refute the role of genetics in human behavior, yet it is written by the same geneticist who was quoted as saying, "Nothing we can know about the genetics of human behavior can have any implications for human society." This statement is no longer compelling, and it clearly illustrates the difficulty that many scientists had in moving beyond the environmentalist doctrine.

What Is Intelligence?

Everyone has an intuitive feel for which of their friends is most intelligent, but a concise and unambiguous definition of the quality is very difficult to achieve. The philosopher Homer believed that intelligence is a gift of grace that not all men possess. Many centuries later, in 1923, with far less insight than Homer could muster, E. G. Boring claimed that "intelligence is what the intelligence test measures." This circularity of definition represents very nearly the state of the art today; we cannot define the quality well, yet we claim to recognize it in others, and to measure it accurately with some fairly simple tests. Nevertheless, whatever trait is actually measured by an intelligence test does tend to be rather constant over one's lifetime, and does have some ability to predict success in school and in the workplace.

What seems clear is that intelligence is not a trait like height, with a single dimension that is easy to measure.[38] Virtually all scientists who study intelligence agree that intelligence involves the exchange of information between working memory and long-term memory. In computer terms, this is analogous to exchanges between random-access memory (RAM) and the hard disk. In simpler terms, this can be thought of as exchange between your desk top and your desk drawers; the information you need right away is kept on your desk top, while the information that is perhaps less pressing is kept in the drawer. All of this information is, of course, updated frequently on the basis of new stimuli in the sensory environment; this is what makes it possible to learn from experience. Thinking is thus the exchange of information between sensory input, working memory, and long-term memory. Modern intelligence tests use several different subtests to measure these exchanges, including word knowledge, short-term memory, deductive reasoning, and the ability to perceive and manipulate patterns implicit in a geometric design. The intelligence quotient (IQ) with which everyone is familiar is actually a weighted composite of the various subtests on an intelligence test. This weighted composite score is normalized to the age of the person tested, so that IQ is actually a ratio of mental age to chronological age (multiplied by 100).

The subtests on an intelligence test are interrelated in a subtle way; even though each was originally developed to measure a different cognitive function, people who do well on one subtest tend to do well on other subtests. People who are gifted in terms of verbal ability tend also to be above average in other mental abilities, such as the ability to manipulate visual patterns or the ability to retrieve things from short-term memory. This suggests that a relatively small number of general abilities can determine performance on what are ostensibly different subtests. Because IQ is calculated as a weighted composite of various subtests, IQ is related to the underlying general mental ability. This underlying "general mental ability" is often called g, which has the advantage that it is far less incendiary than IQ, even though it is functionally equivalent. Yet g is also a statistical abstraction, rather than a direct measure of a definable mental ability.

The general factor g is defined as that component of mental ability that is common to all intelligence tests.[39] Every reliable test of mental ability measures g to some extent, although the degree of correlation with g can vary. The tests that best measure g involve complex cognitive tasks. Tests that are less complex are less able to measure g because they tend to involve simple mental tasks such as sensory discrimination, reaction time, or rote memory. The g factor is of interest because it is not a measure of a specific knowledge, skill, or strategy, but rather reflects individual differences in the speed of information processing. In fact, some people believe that g is actually a measure of a physiological process, such as the speed of conduction of nerve impulses in the brain. The extent to which IQ tests are worthwhile is thus determined by only two considerations:

(1) How well does a specific test measure g? and (2) How well does g determine actual performance at school or at work?

The different subtests on an intelligence exam seek to characterize three basic sets of skills.[38] Verbalization skills are concerned with vocabulary, word use, paragraph comprehension, and so on, while visualization skills involve the mental manipulation of visual patterns. A third category of abstraction skills pertain to reasoning, problem-solving ability, and the ability to find and complete a pattern implicit in a series of related objects. These three sets of skills are clearly complementary to one another, and all are more or less related to g. In a sense, measuring these separate abilities in order to characterize g is somewhat like measuring arm and leg strength to assess muscular strength. A person can have arm or leg function impaired by some factor unrelated to muscle strength, but usually arm and leg strength are a good indicator of general muscular strength. But, possessing great muscular strength will not, in and of itself, make someone an athlete, just as having a high IQ will not guarantee success in life.

Intelligence is perhaps best defined as the ability to solve problems quickly and efficiently. Life often seems to be an endless series of problems, so a great premium is placed on an ability to solve these problems in the time allotted. To achieve great success in life, a person must possess at least a modicum of intelligence, but that is clearly not enough. Intelligence must be complemented by perseverance, self-confidence, and energy, and great intelligence often cannot overcome poor health, laziness, poor social skills, or a lack of initiative.

Can Intelligence Be Accurately Measured?

If intelligence is defined as the ability to solve problems, then in principle it should be possible to measure intelligence using a carefully posed set of problems. This is the basic rationale for all mental testing, and most psychologists argue that current mental tests are capable of measuring intelligence with acceptable accuracy. In fact, a letter published recently in *The Wall Street Journal* and signed by 52 of the most prominent experts in intelligence[40] stated that "intelligence . . . can be measured, and intelligence tests measure it well. They are among the most accurate (in technical terms, reliable and valid) of all psychological tests and assessments." The authors added that "while there are different types of intelligence tests, they all measure the same intelligence." There are some dissenters from this majority viewpoint, but the important point is that consensus has largely been achieved.

One of the best-known dissenters is Howard Gardner of Harvard University, who argues that there is no general mental ability g that can be measured by a test of logical thinking.[37] Instead, Gardner argues that there are seven distinct types of intelligence: linguistic, musical, spatial, logical-mathematical, bodily-kinesthetic, intrapersonal, and interpersonal. Critics have argued that this simply broadens the definition of intelligence to include what might more properly be called talents, but Gardner responds that language and logical thinking are also just talents. Gardner's argument is somewhat appealing, but the majority of scholars now endorse g as adequate to describe intelligence.

A fascinating feature of the human brain is that neurologic damage can essentially delete certain mental abilities. Stroke, brain injury, or the growth of a brain tumor can produce damage in a small portion of the brain, and such damage can cause a person to lose a small portion of his normal repertoire of behavior. Oliver Sachs, the neurologist who wrote *The Man Who Mistook His Wife for a Hat*, has reported several bizarre neurologic syndromes, of interest not because they are common, but because of what they reveal about the workings of the normal human brain. For example, Sachs describes a patient who is wonderfully articulate but cannot think of the name of any common objects; once the object is named, though, he is able to recognize it and use it as would a normal person. Another patient has no recollection whatsoever of the last 40 years of his life, although he remembers the first 20 years of his life in vivid detail. Yet another patient is unable to understand the meaning of words, but is able to read the subtext of speech, written on the face of the speaker, with uncanny accuracy. These bizarre deficits demonstrate that abilities such as word recollection, memory, and verbal comprehension have distinct loci within the brain, and that it is possible to lose an ability without being otherwise impaired. These deficits also suggest that different mental abilities, which are ordinarily very closely related to one another, can sometimes come "uncoupled" in an individual patient, as a result of structural brain damage. These considerations are not really relevant in the vast majority of people, but they do imply that we perhaps have a simplistic definition of intelligence.

Does Intelligence Correlate with Performance?

If our current definition of intelligence is perhaps flawed, we must ask the question, to what extent does measured intelligence predict success in life? Perhaps the best source of data relating to this question is the National Longitudinal Survey of Youth, a study organized by scientists at the University of Chicago.[41] This study has followed more than 10,000 children for up to 27 years, measuring intellectual and socioeconomic variables all the while, and it is widely recognized as the best longitudinal data in the country. The study, which originated under the patronage of the Bureau of Labor Statistics in the mid-1960s, is actually several separate longitudinal surveys, some of which have been discontinued. The longest continuing study was begun in 1968, and includes a sample of women then between the ages of 14 and 24. In 1979, another group of 15,000 young people between the ages of 14 and 21 was surveyed; this group is known as NLSY79 and is the study group used in *The Bell Curve* analysis. Although budget constraints in the early 1980s caused the study population to be reduced to 10,000, these participants are still active in the study.

At the inception of the NLSY79 study, participants were surveyed to collect a wide range of data on attitudes and demographics. Each participant completed questionnaires relating to parental socioeconomic status (SES), childhood environment, and religious beliefs, and each person also took the Armed Forces Qualification Test, a widely accepted measure of IQ. These participants have been interviewed annually, as their life unfolds, and now more than 7000 children, born to the original participants, are also enrolled in the study. This data base is an absolute gold mine of information; more than 2400 books, newspaper articles, and dissertations have been written about the study participants, and the survey will become progressively more valuable with time as study participants age.

The NLSY79 data base was analyzed by Richard Herrnstein and Charles Murray as a part of *The Bell Curve*,[37] to determine the relationship between IQ and various measures of success in life. Low IQ can be thought of as a "risk factor," or a factor that predisposes someone to a risk such as failure to finish high school. (For our purposes, a person of below-average intelligence is defined as someone with an IQ from 75 to 90, while a person of above-average intelligence is defined as someone with an IQ from 110 to 125.) It is then possible to calculate whether a person with low IQ is more vulnerable to the vagaries of life using the following method. If a person of below-average intelligence has a 10% chance of engaging in some particular behavior, while a person of above-average intelligence has a 2% chance of engaging in the same behavior, then low IQ is associated with a 5-fold increase in relative risk of that behavior (10% divided by 2% = 5). In this way, the relative risk of various misfortunes can be calculated as a function of IQ.

The NLSY79 data base shows that poverty is more than 5fold as common among whites of below-average intelligence, compared to whites of above-average intelligence. Compared to a white person of above-average intelligence, a white person of below-average intelligence is: 70-fold more likely to drop out of school without obtaining a high school diploma; eightfold more likely to go on welfare; seven-fold more likely to go to jail; fivefold more likely to live in poverty; fivefold more likely to raise children in an unsatisfactory environment; fourfold more likely to bear illegitimate children; twice as likely to have children with major behavioral problems; twice as likely to be job-disabled; and almost twice as likely to be unemployed.[37] It could well be argued that most of these misfortunes are predicated on the inability to obtain a high school diploma, but this does not change the fact that a person of below-average intelligence is 70-fold more likely to drop out and thereby put himself at risk for all of these contingent problems.

These findings are not unique; a good deal of evidence exists that high IQ is associated with greater success in life. For example, a study in Norway examined both identical and fraternal twins, to determine whether IQ is correlated with educational attainment or occupational status.[42] It was found that the correlation between IQ and education was 0.52, meaning that about 25% of the variability in educational status could be explained on the basis of IQ alone. Similarly, the correlation between IQ and job status was 0.33, meaning that at least 10% of the variability in job status could be explained by IQ alone. While 10% may not seem like much, it should be remembered that this ignores the contribution of health, education, and hard work to job status. On the basis of these results, it was calculated that the heritability of IQ is about 66%, while the heritability of educational level is 51%, and the heritability of SES is 43%. Scientists thus concluded that IQ is largely responsible for both educational attainment and job status.

The Genetics of Intelligence

Deciphering the genetics of intelligence is difficult because intelligence is so all-pervasive, both in defining our world view and in structuring our society. Rightly or wrongly, a person tends to be sorted on the basis of intelligence early in life, so that the majority of people one encounters on a day-to-day basis are of roughly comparable intelligence. This self-sorting and sorting-by-society extends from early school-age to old age, from self-selected classes in junior high school to occupational roles in adulthood, from one's choice of a spouse to one's circle of friends. Because human society is, to a certain extent, structured by intelligence, some of the assumptions that are routinely made by geneticists trying to decipher the heritability of a trait are violated. For example, virtually all of our understanding of genetics is based on the idea that mating between individuals is random. Yet we know that intelligence is often used as a basis for selecting a spouse. There is usually a very good correlation between the IQ of spouses, so this necessarily means that the assumption of "assortative mating" is violated, at least for IQ.[43]

Understanding the genetics of intelligence is complicated by the fact that intelligence is apparently not inherited as a unit, so that it is possible to inherit different aspects of the intelligence of your parents. For example, verbal and spatial memory are as strongly heritable as is overall IQ, but there is evidence that memory may be considerably less heritable.[43] Deciphering the inheritance of a complex trait like intelligence is doomed to almost certain failure when using ideas developed from analyzing the inheritance of simple traits. Yet the complexity of human intelligence may be less of a problem than it seems at first, because many aspects of intelligence are inadequately assessed by intelligence tests. If a particular test is unable to identify someone who is very gifted musically, then the test will also miss this ability in all other subjects. Therefore, while our understanding of intelligence may be too simplistic, this understanding is probably comparably simplistic for all people. A major problem would occur only if a particular intelligence test was able to measure ability in some, but not all, subjects.

Many genes are apparently involved in determining intelligence, so the specific contribution of any one gene is rather small.[44] A recent study of children with low, medium, and high IQ has suggested that there is at least one major gene involved in producing great intelligence, but this idea is really quite speculative at present.[22] In truth, no one has the slightest idea how many genes are involved in producing a person of great intelligence. But it is a fair bet that someone of moderate intelligence is produced by relatively fewer favorable genes than is someone of great intelligence. While each gene may have a small impact on intelligence, the cumulative effect on IQ could be quite large.

Finally, the expression of intelligence is very much at the whim of external circumstances, and even somewhat at the mercy of internal circumstances. As we have seen, childhood exposure to lead in the environment can lower intelligence in even the most supportive intellectual circumstances. Since most intellectual circumstances are less than completely supportive, it is quite likely that the environment routinely conspires to produce a lower intelligence than is specified by the genes alone. And different genes interact with each other in ways that are far from predictable. One gene may be dominant to another, or to a whole series of other genes, while certain genes may subtly modify the expression of even a dominant gene.

As an example of the genetic complexity of intelligence, there is evidence that mathematical intelligence is, to at least some extent, sex-linked.[45] This means that, in an unknown way, genes that determine an individual's sex interact with genes associated with the ability to do mathematics. This was shown clearly in a study of 9927 intellectually gifted junior high school students, each of whom took the Scholastic Aptitude Test, a test that is normally intended for college-bound juniors and seniors. It was found that boys scored, on average, about 40 points higher than girls on the mathematical part of the test, even though boys and girls, at that point, had the same amount of formal training in mathematics. In other words, a substantial difference in mathematical ability existed between boys and girls before this difference could be attributed to different courses of study. The differences between boys and girls in mathematical ability could not be attributed to overall differences in intelligence, since boys and girls scored equally well on the verbal part of the examination. It is also unlikely that this difference was related to differences in environment or socialization prior to testing, since no such differences could be found. Instead, it was concluded that sex differences in mathematical achievement result from an innately superior mathematical ability in males; this is consistent with a previously known greater ability of males in spatial tasks.

Table 2
Average Correlation of IQ in Families[a]

Relationship	Correlation	No. of pairs	Range of correlations
Same person tested twice	90%	88	NA
Identical twins reared together	86%	4,672	58–95%
Identical twins reared apart	72%	110	69–75%
Fraternal twins reared together	60%	5,546	20–87%
Fraternal twins reared apart	52%	34	NA
Siblings reared together	47%	26,473	11–90%
Siblings reared apart	24%	203	23–25%
Parent and offspring reared together	42%	8,433	5–87%
Parent and offspring reared apart	22%	814	9–38%
Adoptive siblings reared together	29%	345	5–38%
Spouses	33%	3,817	16–74%

[a]Data from 111 separate studies of intelligence, involving more than 113,942 pairwise comparisons (e.g., identical twins to each other or mother to daughter), collated from the literature,[44] and supplemented where necessary.[51] Heritability calculated from these data are: by the direct method, 72% (i.e., identical twins reared apart); or by the indirect method, 52% [i.e., twice the difference between identical twins reared together (86%) and fraternal twins reared together (60%)]. The average correlation between spouses for intelligence is higher than for almost all other traits; this indicates that IQ is responsible for assortative mating and suggests that first-degree relatives are more similar for IQ than for most other traits. It is of interest that fraternal twins are no more similar to each other genetically than are full siblings, even though the IQ of fraternal twins is much more similar.

between twins is equivalent to the heritability of IQ, since there is no "shared environment." By this logic, the heritability of intelligence is 74%,[8] since intelligence was 74% similar between identical twins in a sample of 110 twin pairs. In other words, 74% of the variation in intelligence of identical twins could be explained on the basis of genes alone, without any contribution from the environment. This means, of course, that the remaining 26% of the variation in intelligence must be a result of environmental factors. This estimate sets the upper bound of plausibility, because identical twins reared apart do share some amount of environment, even if they are reared apart. This is because identical twins reared apart are free to construct their own environment to a certain extent, and the choice of personal environment is influenced by the genes. In addition, children born at the same historical moment are very likely to share elements of the environment external to the home, as noted earlier. The bleakness of the Great Depression, the hopefulness of the era of space exploration in the 1960s, and the unconstrained greed of the 1980s all must have some effect on a child. Therefore, we regard the actual heritability of intelligence as something less than 74%.

The indirect method of determining heritability relies on determining the correlation between identical twins reared together and fraternal twins reared together (Table 2). Heritability is then calculated by subtracting the IQ correlation for fraternal twins from the IQ correlation for identical twins, and doubling the result. By this logic, the heritability of intelligence is about 52%.[22,43] But the IQ of fraternal twins tends to be more closely correlated than the IQ of other (nontwin) siblings, even though fraternal twins are, genetically speaking, no more closely related than other siblings. This implies that there is more "shared environment" for fraternal twins than for other siblings, and that fraternal twin environment is more concordant than normal. Thus, the indirect method of calculating heritability may underestimate the influence of this "shared environment." There is an uncertainty built into the indirect estimate of heritability, meaning that true heritability must fall somewhere within the range of 30 to 70%, but it is problematic to determine heritability any more exactly.

There is also reasonably good evidence that heritability varies with age.[46] The Colorado Adoption Project, which was initiated in 1975, tested the intelligence of biological parents and their children given up for adoption, and it also tested the intelligence of adoptive parents. Children were tested repeatedly at 1, 2, 3, 4, and 7 years of age, and the correlation between children and their biological and adoptive parents was analyzed. From the correlation between children and their biological parents, it was calculated that the heritability of intelligence was only 9% at age 1. But heritability increased progressively with increasing age, as the genes were given time to assert themselves. By age 4, heritability was 20%, and reached 36% at age 7. Studies that examined older children imply that heritability reaches 45–51% by late adolescence,[47] and may reach 80% late in life.[2] Of course, these findings may simply mean that the tests given to very young children are flawed; as children grow older, the tests used are less flawed, so that the intellectual similarities which were always there just become more apparent. Nevertheless, this study suggests that both genes and the environment are important in determining intelligence in the teenage years, but that genes can make an additional contribution to intelligence as children grow older.[47] This confusing picture does not make it any easier to arrive at a simple estimation of the heritability of intelligence.

Furthermore, the heritability of intelligence may differ by intelligence level. One study that examined identical and fraternal twins concluded that the heritability of intelligence was higher for high IQ, meaning that intelligence is more nearly hereditary than is stupidity.[48] But another study concluded exactly the opposite, that the heritability of IQ is greater for low IQ.[43] This seems to be somewhat easier to rationalize, as it implies that one can inherit a vulnerability to low intelligence, much as one can inherit a vulnerability to Alzheimer's disease. All we can do at this point is to restrict ourselves to determining average heritability of the average IQ at an average age of about 30.

Mathematical analysis suggests that heritability will consistently differ when calculated by the indirect and direct methods.[43] In general, heritability calculated by the direct method tends to be higher than that calculated by the indirect method.

The differences do not appear to be related to such factors as selective placement of adoptees with adoptive parents of similar intelligence, or other interactions between genes and the environment. Instead, analysis suggests that the subtle similarity of environmental factors within a family is important in determining intelligence.

Given these uncertainties, mathematical models of inheritance may be needed to obtain a good estimate of the heritability of intelligence. A great deal of effort has been devoted to making such models, and to incorporating new data into the models, to get a good estimate of heritability. To make a long story short, a current mathematical model estimates that the heritability of intelligence is between 47 and 58%.[27] Another similar model has arrived at an estimate of between 54 and 64%.[49] On balance, we take the average heritability of intelligence to be 60%; this estimate is necessarily tentative, but it is likely to be fairly conservative for a mature adult. This means, of course, that 40% (or more than one-third) of the average intelligence can be attributed to differences in environment. This must mean that the environment is critically important in determining intelligence; a potentially brilliant child in a depauperate environment may lose all hope of brilliance, while an average child, lacking an adequate education, may become functionally well below average.

Heredity and Environment Interact Strongly

Mathematical models of the heritability of intelligence suggest that environmental factors within the family are important in determining intelligence. This conclusion, as simple as it seems, has never been thoroughly tested by scientists. There is really only one study that tries rigorously to partition the impact of heredity and environment on IQ.[50] This study used a very ingenious study plan (known as a cross-fostering design), to show that heredity and environment are each critical in determining IQ (Table 3).

Table 3
IQ of Adopted Children[a]

		SES of adoptive parents		
		High	Low	Average
SES of biological parents	High	120	108	114
	Low	104	92	98
	Average	112	100	

[a]Data from a cross-fostering study involving a total of 38 children.[50] Effects on IQ related to socioeconomic status (SES) of both the biological and adoptive parents are very significant, although the effect of the biological parents SES was more significant.

This important study set out to assess the effect of socio-economic status (SES) on IQ, by determining how the IQ of adopted children is affected by the SES of their parents.[50] Rigid criteria were used to identify study cases, born in France between 1970 and 1975, who were given up for adoption when quite young. The SES of biological and adoptive parents was scored objectively as being either high, medium, or low, then all parents of medium SES were excluded from further consideration. This was done simply to contrast extremes, to maximize the chance of being able to discern a role for SES. Adoption records were sought specifically for low-SES children who were adopted by low- and high-SES parents, in order to determine whether exposure to a high-SES environment could increase IQ. Similarly, records were sought for high-SES children who had been adopted by low- and high-SES parents, to determine whether exposure to a low-SES environment could reduce IQ. More than 600 adoption records were reviewed, to identify ten adopted children who fit into each of these four rigid types. Then the IQ of each of these adopted children was tested.

Overall, it was found that the IQ of children born to high-SES parents is almost 16 points higher than that of children born to low-SES parents (Table 3). Conversely, the IQ of children adopted by high-SES parents is about 12 points higher than that of children adopted by low-SES parents. Obviously, it is best to be born to parents of high SES and then adopted by parents of high SES; the average IQ of children who fit this profile is an impressive 120. It is worst to be born to parents of low SES and then adopted by parents of low SES; the average IQ of children who fit this profile is only 92. But a child born to low-SES parents and adopted by high-SES parents could count on having an IQ more than 11 points higher than a similar child adopted by low-SES parents. Similarly, children born to high-SES parents who were adopted by high-SES parents could count on having an IQ more than 12 points higher than a similar child adopted by low-SES parents. What all of this means is that both genes and the environment are critical in producing an intelligent child. On average, high SES of the biological parents was sufficient to raise the IQ of the children by nearly 16 points. But high SES of the adoptive parents was sufficient to raise the IQ of the children by about 12 points. As we shall see at the conclusion of this chapter, a difference of 12 IQ points can make a tremendous difference in the quality of life.

This is a wonderfully clear demonstration of what we have always intuitively known: both genes and the environment must be adequate to produce an intelligent child. These results could not be explained on the basis of differences in health of the children, since the birth weight, the length of pregnancy, and the prevalence of newborn illnesses were comparable among the four groups. The fact that such a clear result could be obtained without studying identical and fraternal twins is also refreshing; one begins to suspect that it is possible to overdo twin studies. From a genetic standpoint, it is also interesting that the effect of high SES of the adoptive parents was very similar for children born to either high- or low-SES parents; in both cases, high SES of the adoptive parents raised IQ by roughly 12 points. This warrants the simple conclusion that SES alone can account for an approximate 12-point swing in IQ. In this context, it is very interesting that *The Bell Curve* concludes that the average IQ difference between blacks and whites is 15 points. It is perhaps no coincidence at all that the racial difference in IQ is almost identical to the SES difference in IQ.

How *The Bell Curve* Is Wrong

The Bell Curve contends that intelligence is a critical factor in explaining many features of our society beyond simply educational success and occupation. Intelligence is also invoked to explain features of society such as rates of divorce, illegitimacy, unemployment, welfare dependency, poverty, and crime. It is argued that various ethnic groups differ in average intelligence, and that intelligence is an evermore-critical factor in structuring American society. Data are abstracted from the National Longitudinal Survey of Youth (NLSY) to show that there is a relationship between intelligence and the patterns of life in America. The authors then describe social programs they believe to be justified on the basis of the importance of intelligence in structuring society. They contend that intelligence cannot be increased significantly by environmental improvement, saying that "... the story of attempts to raise intelligence is one of high hopes, flamboyant claims, and disappointing results. For the foreseeable future, the problems of low cognitive ability are not going to be solved by outside interventions to make children smarter" (p. 389). Because they perceive intelligence to be an immutable property of the individual, social stratification on the basis of intelligence is seen as inevitable, perhaps even beneficial, for American society. The authors conclude by arguing strongly against affirmative action, in colleges and in the workplace, because they believe it to be "leaking a poison into the American soul" (p. 508).

This book has been given a poor reception by most members of the intellectual community. Many negative reviews of the book have appeared, but these reviews are often flawed in the same way that *The Bell Curve* itself is flawed. Both the book, and most reviews of it, are written as polemics, in the sense that they are concerned more with bludgeoning an alternative viewpoint than with discerning the truth. In fact, the early part of *The Bell Curve*, where data from the NLSY are analyzed, does a better job of marshalling evidence than a good many critics were able to do. The most critical reviews are often very poorly done; some of them are simply an emotional denial of the book, without a serious consideration of its content, while others are the result of reflexive and somewhat myopic liberalism, not intellectual criticism.

Nevertheless, *The Bell Curve* is very seriously flawed; while the data from the NLSY seem, for the most part, to be well analyzed, there is still a huge leap involved in going from these facts to the interpretation given them. As a letter in *The Wall Street Journal*, signed by 52 leading experts in intelligence, notes, ". . . research findings neither dictate nor preclude any particular social policy, because they can never determine our goals."[40] In other words, research findings are (if we are lucky) facts, while social policies can be no more than interpretations given to those facts. This distinction between fact and interpretation is a very critical one, especially for a scientist; basically, facts don't change, but the interpretation given to them can change radically. A scientist must be extremely careful in going beyond the data to synthesize and generalize, and even more careful when presenting generalizations to a lay audience that is perhaps less aware of the distinction between fact and interpretation. Many scientists would argue that it is inappropriate for someone trained in science to make pronouncements about public policy at all. However, we believe that if scientists fail to make public policy recommendations based on good science, then politicians will make them based on bad science.

In short, *The Bell Curve* would be a better book if it had three parts, each clearly separated from the other, and each designed to answer a single question, as follows:

1. What is the pattern? What is the evidence that intelligence is correlated with other key societal variables, including divorce, illegitimacy, unemployment, welfare dependency, poverty, and crime? *The Bell Curve* actually does quite a good job answering this question.
2. How was this pattern established? If intelligence is found to be important in structuring American life, one

must address the genesis of intelligence. If nature alone determines intelligence, then a defeatist tone is perhaps appropriate, but if nurture also determines intelligence, then giving up on the disadvantaged is totally unjustified. The book fails to address the genesis of intelligence adequately.

3. What should we do about the pattern? This is clearly a public policy issue, so this section of the book would necessarily be interpretative, but it should also be firmly grounded in facts collated in the preceding two parts, and interpretations should be clearly acknowledged as such. The book also fails to answer this question, because it freely mingles fact and interpretation in a very volatile fashion.

In our estimation, *The Bell Curve* is seriously flawed in two ways: it first misconstrues a very critical fact, and it then fails completely to indicate where fact stops and interpretation begins. The critical fact misconstrued in *The Bell Curve* is the meaning and importance of heritability. The authors first acknowledge that intelligence is only about 60% heritable, but they later state, as quoted above, that intelligence is basically immutable. A heritability of 60% means that genes are more important than environment in determining intelligence, but it also means that environment absolutely does have a role in determining intelligence. Genetics may determine the range through which environment can modify a trait such as intelligence, but genetics cannot preclude the environment from having a very potent impact on intelligence. To conclude that a trait that is only 60% heritable is also immutable is a very grave error. It is an especially critical error because the authors prescribe a social program based on their naive conception of heritability.

The authors of *The Bell Curve* state openly that the heritability of intelligence is 60%, but through much of their discussion they seem to tacitly assume that heritability is actually closer to 80% or even 100%. Unless one assumes that DNA is destiny, how can one rationalize the statement that "formal schooling offers little hope of narrowing cognitive inequality" or that "the problems of low cognitive ability are not going to be solved by outside interventions to make children smarter" (p. 389)? The authors also say that "the more one knows about the evidence, the harder it is to be optimistic about prospects in the near future for raising the scores of the people who are most disadvantaged by their low scores" (p. 390). There is a message of profound pessimism here that has not been lost on the media. Yet just because we have failed in the past to raise intelligence by environmental enrichment does not mean that we will always fail. Clearly, there is a role for genetics in establishing the patterns noted in *The Bell Curve*, but just as clearly, the pattern was established and is maintained by systematic discrimination against the poor and disenfranchised.

A seemingly small change in average IQ of a group can produce rather radical changes in social behavior, according to no less an authority than *The Bell Curve* (p. 368). The mean IQ of children in the National Longitudinal Survey of Youth was about 100, as it is expected to be in any large random sample of people in the United States. If this mean is altered by only 3 points, to 97, by randomly excluding some of those people who pull the average IQ up to 100, then there are a number of striking changes in social problems. As the IQ falls by only 3 points, the number of women ever on welfare increases by about 13%, the number of men ever incarcerated increases by about 11%, the number of people below the poverty line increases by about 10%, and the number of children born out of wedlock increases by about 7%. If a similar manipulation is done to raise IQ by only 3 points, to 103, by excluding those people who pull the average down to 100, there is a striking reversal of these social problems. As the IQ increases by only 3 points, the number of women ever on welfare decreases by about 19%, the number of men ever incarcerated decreases by about 25%, the number of people below the poverty line decreases by about 26%, and the number of children born out of wedlock decreases by about 17%. Is it any wonder, then, that the 15-point IQ difference between blacks and whites, so forcefully noted by *The Bell Curve*, is associated with grave social problems which impact the black community?

In reality, the problems epidemic in the ghetto are likely not racial in origin, and they cannot be lessened by "solutions" that increase racial disparity in SES. Disparity in social environments could easily account for the IQ differences described in *The Bell Curve*, and they could as easily account for differences in social problems such as drug use, alcoholism, and violence. Official policy that would perpetuate these inequities is simply lighting a fuse to an already explosive situation. The social programs suggested in *The Bell Curve* are intellectually bankrupt and morally indefensible; they resolve nothing, they merely validate an inclination to ignore the problem. By advocating benign neglect as an appropriate response to poverty and social disenfranchisement, Herrnstein and Murray have simply given up on the problem, and thereby done immense harm to a large number of people. But great minds don't give up on a problem just because the problem is difficult.

As an ironic aside, in a perfect world where environment is equally enabling for everyone, the heritability of intelligence would actually increase. If differences in environment do not exist, then genetics must explain 100% of the IQ differences among people. Thus, we do not imagine that all obstacles to achievement for all people will be removed by making a world where the environment is equally enabling for everyone. Instead, intelligence must be recognized for what it is: a gift of grace. Those blessed with intelligence should work very hard to make a better world for all, because inherent in the belief that intelligence is a gift is the idea that the gift bears certain responsibilities.

9

Mental Disorders

Evidence abounds that intelligence is heritable, which of course implies that a range of other mental states are also heritable. Mental illness was known to run in families centuries ago, and both manic-depression and schizophrenia were recognized as familial diseases by Sigmund Freud. But progress in defining the genetic basis for mental illness has been slow, in part because a definitive diagnosis of mental illness is often so hard to make. In fact, research into the behavioral genetics of mental illness has been filled with false starts, misleading data, and dashed hopes.

However, there is more reason now than ever before to be optimistic that the genetic basis of mental illness can be deciphered. Mental illness is understood in a way fundamentally different now than before; we know that mental illness is a medical problem, not demonic possession or divine visitation. Psychoanalytic tools have become sophisticated enough that psychiatrists can achieve consensus on most diagnoses. While this may sound like a small achievement, it represents a fairly profound advance over the situation of just a generation ago.

Mental Retardation Is Strongly Heritable

Mild retardation is probably only the lower end of a broad spectrum of intelligence, whereas severe retardation may be a genetically more distinct entity. The genes involved in determining high or low intelligence appear to be quite distinct from the genes that can cause severe mental retardation. Fragile X syndrome is an example of a form of mental retardation that appears to involve genes distinct from those normally involved in the determination of intelligence. This genetic syndrome can cause profound mental retardation, as a result of an abnormal gene carried on the X chromosome. Fragile X syndrome affects boys more frequently than girls, because males have only one X chromosome, whereas females have two; thus, a mutation that affects only one X chromosome can be compensated for by a normal gene in females, but not in males. In fact, among children with the fragile X mutation, 100% of affected boys have some degree of mental retardation, while only half the girls are mentally retarded.[52] Severe or profound retardation is more than eightfold more common in fragile X boys than in fragile X girls. Since this mutation is carried by about 1% of the Caucasian population, fragile X syndrome is one of the more common causes of mental retardation.

It is of interest that fragile X syndrome is associated with behavioral disturbances as well as mental retardation. Again, boys with fragile X syndrome are nearly sixfold more likely to suffer moderate to severe behavioral disturbance than are girls afflicted with fragile X syndrome. Fragile X syndrome is strongly heritable, but there is also speculation that the environment can play some role in determining the severity of disease. This speculation arises because the fragile X mutation can cause any level of mental retardation, from borderline to profound. In addition, the mutation can apparently assume a more severe form in an individual after conception. This implies that some feature of the environment, perhaps experienced by the mother while the affected child is *in utero,* may in some way determine the severity of retardation.

Our knowledge of fragile X syndrome demonstrates very clearly that mental retardation can be hereditary. Nevertheless, we will not focus here on mental retardation, because the most recent progress in behavioral genetics has related to other mental disorders. Instead, having shown strong evidence that mental disorder can be inherited, we will focus our attention on manic-depressive illness and schizophrenia, because these mental illnesses have been the focus of so much recent research. In addition, we will examine several other mental illnesses, which seem to reveal some of the underpinnings of human behavior.

Inheritance of Manic-Depression

Manic-depressive illness, also known as bipolar affective disorder, is familiar to many because a list of manic-depressives is a Who's Who of the Arts. The list includes Walt Whitman, Vincent van Gogh, Edgar Allan Poe, William Blake, Ernest Hemingway, Virginia Woolf, Winston Churchill, Gustav Mahler, Cole Porter, Sylvia Plath, Lord Byron, Georgia O'Keeffe, Jack Kerouac, Tennessee Williams, Alfred Lord Tennyson, and Mark Twain. But this list of notables is somewhat misleading, in that it seems to confirm the stereotype that manic-depression somehow causes creativity. In fact, most sufferers of the disease are barely able to deal with the devastating consequences of their disease, and are unable to scale the creative heights achieved by these notable few. Manic-depression afflicts roughly 7 in 1000 people, so the lifetime risk of manic-depression is less than 1%. But, for those who have a parent or sibling with manic-depression, the risk of disease is about tenfold higher.

Manic-depression is a cyclic alteration of mood, from clinically depressed to wildly elated, with mood swings that can be moderate to very severe, and can occur over a highly variable time frame. The depressive phase of the illness causes an intense melancholy characterized by hopelessness, apathy, anergy (exhaustion), sleep disturbance, slowed physical movement and thinking, impaired memory and concentration, and emotional blunting. Conversely, the euphoric or manic state is associated with hyperactivity and irritability. In the grip of the manic state, a manic-depressive may spend money recklessly, drive dangerously, pursue questionable business deals, and engage in casual sexual relationships. Manics speak rapidly, are easily distracted, have intrusive thoughts, are excitable, move and think very quickly, hold opinions with great conviction, and live life with impetuousness and a certain grandiosity. Advanced cases of manic-depression are marked by dramatic, cyclic shifts between mania and depression, whereas a mild form of the disease, known as cyclothymia, can cause nondebilitating changes in mood, behavior, sleep, thought patterns, and energy levels. The diagnostic criteria for manic-depression include suicidal thinking, self-blame, and inappropriate feelings of guilt in the depressive phase of the disease. In the manic phase, the most severely affected individuals suffer violent agitation, auditory or visual hallucinations, and delusional thinking. Strikingly, half of all suicides in the United States are committed by persons with either manic-depression or clinical depression.

Studies of twins separated at birth, and of adopted and biological children reared together, have shown that manic-depression is more strongly heritable than clinical depression. Yet there is still evidence for an environmental component to manic-depression. Children born before 1940 are at a lower risk of developing manic-depression than children born between 1940 and 1959. In fact, the incidence of manic-depression among the later cohort of children is about as high at age 30 as in the earlier cohort at age 65; this suggests that some feature of the environment is inducing earlier expression of the disease in the younger cohort.[53] In addition, the suicide rate among 15- to 19-year-olds is ten times higher for those born in the late 1950s than for those born in the early 1930s. Since a large fraction of suicides are committed by those suffering from manic-depression, this also suggests that the incidence of manic-depression has increased. However, these differences are potentially the result of better clinical diagnosis or better record-keeping by public health officials. Moreover, no one yet knows what features of the environment play a role in the expression of manic-depression; clearly, adversity can evoke depression, but what life events could possibly evoke the wild mood swings of manic-depression?

Manic-depression is a heritable trait, and many separate studies have shown that identical twins have a higher concordance for the disease than do fraternal twins.[54] Pooled results from several separate studies show that the concordance for identical twins is about 62%, whereas the concordance for fraternal twins is about 18%. An even more revealing fact is that twins tend to share the same type of emotional disorder; in one small study, 79% of identical twins were concordant for clinical depression, while 78% of identical twins were concordant for manic-depression. Finally, a small study of identical twins reared apart showed 67% concordance for manic-depression, even though these twins shared no environment. Thus, there can be no doubt that this form of mental illness is inherited. Yet the exact degree of heritability has been debated endlessly: the low-end estimate of heritability, based on a reasonable sample of twins, is about 67%, whereas the high-end estimate, based on a fairly large sample of twins, is about 88%.[55] It seems clear that manic-depression is more strongly heritable than is clinical depression, and that the more severe the symptoms of manic-depression, the more likely the disease is to be inherited. Yet what is inherited is not the certainty of becoming ill, but merely a susceptibility to illness. This critical insight has led geneticists to conclude that a susceptibility to manic-depression may be present in all of us, but that only certain people pass a threshold that makes them liable to express the disease.

Recent evidence suggests that manic-depression is somehow related to attention deficit disorder (ADD).[56] ADD is known to be familial, as the risk that a child will have ADD is about fivefold higher in families with other afflicted children. ADD can occur with or without hyperactivity, and it affects 6 to 9% of all school-age children. The disorder is a fairly major social problem; the child with ADD is restless, inattentive, impulsive, and easily distracted, so the disorder can cause learning problems for the child and discipline problems for the whole classroom. One study of 73 children with ADD found that more than a third of these children had some type of emotional disorder; in fact, depression afflicted 21% and manic-depression afflicted another 11%. Relatives of the children were also at a higher risk of manic-depression than normal; nearly 25% of the relatives had some type of emotional disorder, whereas only 5% of the general population has an emotional disorder. Thus, the risk of emotional disorder is more than fivefold higher than normal (i.e., 25/5) among close relatives of children with ADD. This is really a striking finding because the current clinical picture of ADD does not include symptoms of mood disturbance. The only reservation about this study is that it examined children who were referred to a clinic for evaluation; one would expect children referred to a clinic to have a particularly severe form of ADD, since most children with ADD are never seen by a health care professional. In any case, these results suggest that there can be a genetic predisposition to both ADD and manic-depression, and some scientists have even suggested that ADD is actually an early symptom of emotional disorder.

Because there is a strong indication that manic-depression is heritable, there has been an intense search for the genes responsible for the disease. In 1987, a great deal of excitement was generated by a report that manic-depressive illness among the Old Order Amish had been linked to a set of gene markers on chromosome 11.[11] This report was exciting for two reasons; first, it meant that we were getting closer to a solid understanding of the disease and that the disease might become more treatable, and second, it meant that genetic testing might soon be able to reveal those most prone to illness. But both of these hopes were dashed when it finally became apparent that the early report was incorrect. Eventually some of the authors of the original study published a second paper[57] retracting the results of the first paper. What went wrong with the study of mania in the Old Order Amish? How were some very convincing results later proven to be totally incorrect? What lessons can be learned from the very public humiliation of a group of well-respected scientists?

Apparently, there were several flaws in the Amish study, unintentional and honestly made, but terribly damaging nonetheless.[11] The Old Order Amish are a group of about 15,000 people, all descended from only 30 pioneer couples who came to the United States more than 250 years ago. The Amish form a perfect group for a study of mental illness: families tend to be large, so that there are many people to study, and family relationships are known for many generations back. Family members all tend to live in the same geographic area, so they are easy to find. Strict prohibitions against alcohol and drug use mean that substance abuse, so frequently linked to mental illness, is not an issue. And finally, the incidence of manic-depressive illness in this group is apparently comparable to other groups, meaning that the genetic mechanism of disease is probably also comparable. The only drawback, apparently not fully realized at the time, is that the close genetic relatedness of these people should mean that their genetic diversity is lower than in a random sample of 5000 people. Thus, any linkage between a gene marker and a disease is likely to be weaker than normal, since the Old Order Amish show a limited range of the genetic diversity likely to be found in another large population. Nevertheless, analysis of a group of 81 people, all members of a single large pedigree of Old Order Amish, showed that 19 men and women were suffering from manic-depression. When the chromosomes from these 19 were studied, using techniques of molecular genetics, it was found that a certain set of gene markers was present on chromosome 11 in all of those persons diagnosed with manic-depression. From the close relationship between the gene markers and the disease, it was calculated that there was only one chance in 10,000 that this linkage could occur by chance.

Yet this extraordinarily close linkage between gene markers and disease fell completely apart when examined more closely.[57] Within 2 years of publication of the first study, a second study was published, which reexamined the first group of 81 people and extended the original sample with information on an additional 39 people. The addition of new people to the analysis had a significant effect, in that several of them had manic-depression without having the same gene markers. This made it look more like the association between the disease and the gene markers was an accident. However, even worse news was in store; psychiatric reevaluation of the people who had been analyzed previously showed that two of them had developed manic-depression after publication of the original study. In fact, adding the new study participants and changing the clinical status of the old participants dramatically changed the odds of linkage; odds that had once been 10,000 to 1 in favor of linkage, were changed to 100,000,000 to 1 against linkage. At about the same time, another study was unable to show linkage between chromosome 11 markers and manic-depression in a second North American pedigree,[58] while a third study failed to find linkage in an Icelandic pedigree.[59] This means either that the gene for manic-depression in Old Order Amish is not located on chromosome 11, or that there are several genes for manic-depression, only one of which is located on chromosome 11. As of yet, it is impossible to tell which of these possibilities is correct.

The misidentification of the gene for manic-depression was a terrible setback for molecular genetics. The mistrust engendered by the Old Order Amish error was compounded when an independent study in Israel was also found to be incorrect.[60] The Israeli study purported to show linkage between manic-depression and gene markers on the X chromosome, and also claimed that there was strong evidence that manic-depression could be caused by a single gene mutation. Now, the best evidence shows no linkage between manic-depression and any gene on the X chromosome.[61] In summary, the location of genes that cause manic-depression is as poorly understood today as it was 10 years ago. While this is a major disappointment, there has been a tendency to discard the baby with the bathwater. We do not yet know where the manic-depression genes are located, but we do know that there are such genes because of the hereditary nature of the illness. While this fragmentary knowledge prevents us from making a blood test for manic-depression, it does not prevent us from continuing a search for the relevant genes.

The case of manic-depression is a particularly poignant example of the potential difficulties that will arise when we have a more complete knowledge of the genetic basis of disease. Even though the symptoms of manic-depression are debilitating and seem terrible to most, there are a substantial number of people with the disease who do not wish to be treated. The reason for their reticence is that, while most would agree that depression is highly unpleasant, many feel that the manic phase of the disease is worth any price. The manic phase, for all its problems, can be associated with unparalleled bursts, even frenzies, of creativity. The composer Robert Schuman was manic-depressive for most of his life,[33] and during his career he wrote 147 musical works formally catalogued as opuses. For two years of his life he was known to be manic or hypomanic and during these two years he wrote 51 opuses. This means that a third of his lifework was written in only two years. In contrast, during four years when he was depressed, he wrote nothing at all. Similarly, Jack Kerouac wrote *The Subterraneans* in three uninterrupted days, and wrote the stage adaptation of *On the Road* in a single night. Evidence is beginning to accumulate that the cognitive style associated with mania or hypomania can lead to a fiery, consuming creativity, characterized by increased fluency of expression and an increased frequency of creative thought.

Scientists have repeatedly shown that there is a link between manic-depression and creativity.[33] An early study of 30 creative writers showed that 80% of them reported experiencing at least one episode of major depression or mania, and 43% reported full-blown mania; this is an incidence nearly ten-fold higher than in the general population. A recent study of 47 distinguished British authors and artists showed that 38% had been treated for mood disorder; mood disorder was especially common among the poets, as half of them had required hospitalization, medication, or both. Another study of 50 modern poets found that 25% of them met diagnostic criteria for depression or manic-depression at the time of the study, and suicide was six-fold more common than normal in this group. About half of the 15 most influential modern abstract-expressionist artists suffered from depression or manic-depression, and the suicide rate in this group is 13-fold higher than the US national rate. A comprehensive analysis of 1005 famous artists, writers, and scientists found that artists and writers experienced two to three times more mood disorder, psychosis, substance abuse, and attempted suicide, than did comparably successful people in business, science, and public life. Again, the poets in this large sample were most prone to mania and psychosis, and their rate of suicide was 18-fold higher than normal.

Given that there is this intriguing link between mania and creativity, and given that many manic-depressives are willing to undergo the torture of depression in order to obtain the high of mania, we must wonder: what is the goal of treating the manic-depressive? We cannot wish to rob the manic-depressive of his creativity, nor the hypomanic of his expansiveness, by flattening out all the cognitive hills and valleys with lithium. Instead, we must strive to find a more effective and compassionate treatment option than is presently available. While we do not want the manic-depressive to be consumed by the fire of creativity, we also do not want the creative fire to be damped out completely in a drug-induced torpidity.

The Familial Nature of Depression

Clinical depression is similar to the depressive phase of manic-depression, without the relief of an alternating phase of mania. For this reason, clinical depression is often called unipolar affective disorder, whereas manic-depression is known as bipolar affective disorder. Very rarely, unipolar affective disorder can cause repeated episodes of mania, without an intervening depressive phase. Clinical depression is distinguished from a normal period of unhappinesss by the continuous presence of symptoms for more than two weeks, at a level of severity that significantly interferes with one's normal function. Clinical depression is much more common than manic-depression, since depression afflicts roughly 11% of people during their lifetime while less than 1% of people are manic-depressive. In fact, the incidence of depression may be much higher than 11%, as one study of a blue-collar neighborhood in London concluded that 70% of women and 40% of men had at least one episode of clinical depression before the age of 65.[54] Overall, roughly 50% of depressed persons recover and never relapse, but 15% never recover; the National Institute of Mental Health estimated in 1981 that nearly 75% of all psychiatric hospitalizations were for clinical depression.

Clearly, clinical depression is often caused by adversity; almost all persons have suffered some sort of setback that caused a period of depression.[62] Those who suffer depression are more likely to have suffered a personal setback in the month immediately before the onset of depression. The difference between clinical depression and normal depression is one of degree, just as setbacks and problems also come in degrees. Therefore, it is reasonable to think that clinical depression may be caused by a major personal setback, such as death of a loved one. However, there is also evidence that clinical depression is genetically linked, and that sometimes depression can come "out of the blue."

One large study found that the risk of clinical depression is higher in certain families than in others, consistent with the idea that depression can be familial.[63] Close relatives of those who are depressed have a near threefold higher lifetime risk of depression than normal, and nearly half the close relatives of a depressed person are likely to have seen a physician about depression at some point in their life. Yet the familiality of depression does not prove that depression is genetic, since families can share depressing life events and depressed attitudes as well as "depressive genes." Split twin studies offer more convincing evidence that clinical depression can be genetic in origin; pooled results of several different studies suggest that about 77% of identical twins are concordant for depression, while only 45% of fraternal twins are so concordant.[64] This suggests that the overall heritability of depression is about 60%. The heritability of depression is higher when depression comes "out of the blue" than when it is induced by some adverse life event. But heritability is necessarily an evanescent concept; it is unlikely to be very accurate for any group other than the one from which the estimate was derived.

Intriguing evidence has been found that some families are more prone than others to suffer adverse life events.[54] Thus, the relationship between depression and adverse life events may arise because certain unfortunate families are prone to both. The frequency of adverse life events is dramatically higher in the relatives of depressed persons than in the general population, even when events related to the depressed person are excluded from consideration. Furthermore, within a given family, the relationship between adversity and depression is weak; there are some individuals who suffer great adversity without suffering any consequent depression, while other individuals may be depressed after relatively minor adverse events. This seems to mean that depression is something that can affect hazard-prone families, not just stress-susceptible individuals. This may mean that a susceptibility to depressive illness and a propensity to adverse life events are different expressions of some more basic heritable trait.

Inheritance of Schizophrenia

Schizophrenia is the most common form of psychosis, affecting about 6 persons in 1000. It is characterized by disordered patterns of thought, including delusions, hallucinations, and withdrawal from social interactions with others, together with an increased investment in an internal world. Schizophrenia is generally not considered to be a single disease entity, but rather a collection of mental disorders with varying symptoms. Cold, clinical terms aside, schizophrenia is usually not the mild dottiness or exaggerated eccentricity it is sometimes portrayed to be; instead, it can be an out-of-control, E-ticket ride through a genuinely frightening mental landscape. Some schizophrenics suffer auditory or visual hallucinations fully as real to them as the external world. Because the schizophrenic is attending to two different worlds, often at cross-purposes with one another, they may be unable to deal effectively with the real world.

Children who are destined to become schizophrenic as adults are likely to suffer from a variety of problems as children, implying that the roots of schizophrenia extend back to early childhood.[65] Children who become schizophrenic generally are not diagnosed until they are 24 years old, on average, but delayed motor development can be present in these children as early as 14 months of age. Children destined to become schizophrenic are likely to prefer solitary play at ages 4 and 6, to have lower intelligence than normal by age 11, to be noticeably less social than normal by age 13, and to be strikingly more anxious than normal by age 15. In addition, mothers of these children are more likely to have a poor understanding of their children and to have relatively weak mothering skills. Yet the presence of these problems in childhood says nothing at all about whether the disease is heritable. Even the association between poor mothering skills and schizophrenia can be explained in two ways: either poor mothering skills induce schizophrenia, or else poor mothering skills reflect a form of hereditary mental illness present in the mother, as well as in the child. In fact, the first of these two possibilities was popular for many years, and the "schizophrenogenic mother" theory only fell out of favor in the late 1960s.

To determine the degree of heritability of schizophrenia, we must again resort to examining identical and fraternal twins. Perhaps the largest study of twins that has been used to address questions of mental illness is the National Academy of Sciences Twin Registry in the United States.[66] This registry was constructed by using a very clever strategy; first, a massive computer search of birth certificates for white male multiple births in the years from 1917 to 1927 in 39 states identified about 54,000 multiple births. Then the names of these people were fed into a computer at the Veterans Administration, which lists all persons who served in any branch of the U.S. Armed Forces. This two-part search revealed 15,924 twins, both of whom had been in the services and both of whom had medical records on file. Identical and fraternal twins were identified, in most cases, by asking the twins themselves whether they were identical. In the relatively few cases where the twins could not respond, the relationship between twins was determined, as well as possible, from medical records. It was then possible to search the medical records for evidence of schizophrenia, as well as for evidence of several other medical conditions.

Unfortunately, this exhaustive procedure identified only 536 sets of twins in which at least one twin was schizophrenic. Of course, this is because schizophrenia is a relatively rare disease, but the relative paucity of affected twins means that any estimate of heritability is necessarily a bit tentative. In any case, about 32% of identical twins shared a diagnosis of schizophrenia, while only 7% of fraternal twins shared a similar diagnosis. Concordance between twins was found to increase with increasing disease severity, meaning that severe cases are more likely to be familial. This should not be too surprising, since someone with a relatively mild degree of schizophrenia may go undiagnosed, while severe cases are more likely to be diagnosed correctly. Based on these concordances, the heritability of schizophrenia is estimated to be about 79%. This estimate is not rock-solid, because the incidence of schizophrenia among these twins was almost twice what is expected in a normal population; this may mean either that schizophrenia was overdiagnosed in this group, or that this group was, for some reason, more prone to schizophrenia. The latter is a very real possibility, since many of these twins must have served in the armed services during wartime, when active duty conditions are likely to induce mental illness in someone who might otherwise only be somewhat unstable. Heritability was also estimated in this group for several physical illnesses, including hypertension, diabetes, heart disease, and ulcers, and it was found that the heritability of schizophrenia was higher than for any of these other diseases. What this means is that schizophrenia is far more likely to be familial than is heart disease, which is probably a surprise to most people.

Concordance for schizophrenia among identical twins tends to be quite high. In addition, there is often concordance for the type of schizophrenia. If identical twins both have schizophrenia, chances are that both twins will have the same type of schizophrenia. Paranoid schizophrenia is less often heritable than is nonparanoid schizophrenia, which implies that paranoid schizophrenia has more of an environmental component than nonparanoid schizophrenia. There is suspicion that a familial susceptibility to schizophrenia must interact with something else in the environment for the disease to be manifest, but what that something else is remains a mystery. Prenatal exposure to the influenza virus is suspected by some epidemiologists, whereas others suspect a type of slow-acting virus, such as the virus that causes *kuru,* or laughing sickness, in the New Guinea Highlands. Stressful life situations seem to evoke schizophrenia, but it is not known whether the "schizophrenic-to-be" participates in creating life stresses.

Recently a great deal of excitement was created by the claim that schizophrenia is linked to a gene on chromosome 5.[67] According to scientists who studied five families in Iceland and another two families in Great Britain, a single gene was responsible for schizophrenia. This gene was supposed to be dominant, meaning that schizophrenia would always be expressed in those who have the gene. This is a stunning conclusion: first, it is surprising that a disorder as complicated as schizophrenia could ever be caused by a single gene; and second, if this is true, it is surprising that the relevant gene was not identified many years ago. For these reasons, many scientists were skeptical that the "schizophrenia gene" had been found. In fact, even scientists involved in the study acknowledged that there might be several different causes of schizophrenia, and they only claimed to have identified one such cause. However, these same scientists also claimed that the gene linked to schizophrenia was linked to several other mental illnesses as well, thereby making their claim even harder to accept at face value.

From a medical standpoint, it would be a great advance if a "schizophrenia gene" could be identified. This would enable psychiatrists to determine who really has the disease, and what form of disease they have, and it might even enable scientists to develop a blood test for schizophrenia, so that people vulnerable to the disease could be identified. But the findings from Iceland do not mean that we now understand the genetics of schizophrenia. The Icelandic study was done in a genetically isolated group of people, in whom random changes could accidentally increase the frequency of a rare set of genes. Furthermore, scientists deliberately tried to increase their ability to find a rare "schizophrenia gene" by studying seven families with an extraordinarily high incidence of disease; among a total of 104 family members, 39 had schizophrenia, 5 had a related mental illness such as schizoid personality disorder, and another 10 had various mental illnesses usually thought to be unrelated to schizophrenia. By concentrating attention on these families with an incredibly high incidence of mental illness, scientists were intentionally trying to bias their sample, to detect genes that are ordinarily quite rare.[68] But it is worth noting that, as rare as schizophrenia is, chances are good that it will cluster in some families purely as a result of bad luck, not bad genes. Although the old adage says that "lightning doesn't strike twice," scientists know that, in fact, it can and does strike twice, or even more than twice.

A firestorm of controversy was generated by this finding of a "schizophrenia gene," and there was a frenzy of activity as other scientists tried to replicate these results. All of the subsequent attempts met with failure. In fact, the same issue of *Nature* that carried the first report also carried a report that was directly contradictory.[69] In the second report, a large Swedish kindred involving 157 men and women was examined, all from only three families. There were 31 schizophrenics among these family members, meaning that the incidence of schizophrenia was more than 30-fold higher than normal. When all individuals were examined, using the same techniques that had been used for the Icelandic kindred, no evidence at all was found for linkage to chromosome 5. In fact, the Swedish cohort showed quite clearly that, at least in this kindred, there was no chance of a linkage to chromosome 5. This means that, even if the Icelandic data are absolutely correct, more than one gene is involved in producing the symptoms known collectively as schizophrenia.

Within 9 months of the first report, it was proven conclusively that the Icelandic data linking schizophrenia to chromosome 5 were actually incorrect.[70] In fact, some of the same scientists who participated in the first study were also involved in the effort to overturn that study. This is a clear example of the self-correcting nature of science; if an error is found in the literature, scientists are very willing to correct that error publicly, because of the prominence such error-correction can bring. Subsequently, other scientists were also unable to show any linkage at all between schizophrenia and chromosome 5.[71] Consensus has now been reached that schizophrenia is not as yet linked to any one gene, and that schizophrenia is more likely to be the result of an interaction between many genes and the environment. Schizophrenia is apparently a very heterogeneous disease, which may explain why it is often difficult to diagnose.

The case for an environmental cause of schizophrenia was greatly strengthened recently by a very striking result obtained using magnetic resonance imaging (MRI). Identical twins who were discordant for schizophrenia were imaged, and it was found that most of these twins were also discordant for brain structure.[72] This means that, even though identical twins have identical genes, they somehow wound up having nonidentical brains, in terms of both structure and function. The most striking difference was that the brain of the schizophrenic twin was almost always smaller than the brain of the normal twin. Findings were so consistent as to suggest that loss of brain mass is somehow characteristic of the disease. But there is an alternative explanation: perhaps loss of brain tissue is somehow a result of treatment for schizophrenia or of malnutrition resulting from long-term schizophrenia. These findings also cannot exclude the possibility that both twins share a genetic tendency to loss of brain mass, but that the tendency was only expressed in the twin who became schizophrenic. Nevertheless, the discrepancy between identical twins does suggest an environmental cause of disease.

Even though twin concordance for schizophrenia is high, it is very difficult to imagine how the disease could be heritable. People afflicted with schizophrenia are often devastated by it, so that they are unable to form loving relationships and so would also be less likely to have children. If schizophrenics have fewer children than normal, then the gene for schizophrenia should disappear from the population over time. The fact that the gene has not disappeared from the human genome must mean one of two things. Either the genetic component of schizophrenia is weak, or those people who carry the gene (but don't express the disease) must enjoy some sort of competitive advantage over other people. This is all very speculative, but perhaps the gene is only expressed in rare circumstances (perhaps following an unusual viral infection?), while in more commonplace circumstances it actually enhances the ability of people in some way. The persistence of some genes in a human population, even though simple competition should eliminate them, is an enduring mystery. How can one explain the long-term existence of a gene for schizophrenia or homosexuality or cystic fibrosis, when people who express these traits are likely to have fewer children than normal?

Inheritance of Alzheimer's Disease

Alzheimer's disease (AD) is a progressive and devastating mental illness that slowly and inexorably causes dementia, thereby robbing millions of people of the ability to care for themselves as they age. AD is often doubly tragic; an independent, intelligent person can be reduced to the indignity of a second childhood, while an undue emotional and financial burden is placed on young people just starting their own family. The disease is associated with death of neurons and an evolving constellation of characteristic abnormalities within the brain. A great difficulty with AD has been that it is a difficult diagnosis to make; older people can lose their sanity for many reasons, and it cannot be assumed that dementia always means Alzheimer's. In fact, a definitive diagnosis of AD can only be made after the death of the patient by finding the characteristic brain lesions at autopsy. However, recent progress makes it very likely that a new diagnostic tool will be available in the near future. This should enable physicians to make a definitive diagnosis soon after symptom onset, and would enable scientists to determine which drugs are best able to arrest disease progression.

AD has been known for a long time as a familial disease, especially in the early onset form, which can affect people in their mid to late 50s. Early onset AD is associated with one of two known mutations, but this was of only marginal interest because early onset AD is so rare.[73] Late-onset AD typically affects people in their mid to late 60s and is about three times as common as early onset AD, although the genetic basis of this disease was unclear until quite recently. But then, in 1993, a group of scientists identified a protein, present in abnormally high concentration in the brain of people with late-onset AD, and our understanding of AD was thus fundamentally altered.[74] This protein turned out to be apolipoprotein E (ApoE), a well-known protein that is normally used to transport cholesterol in the bloodstream. ApoE comes in several different forms, and it turns out that ApoE-4, a relatively uncommon form of ApoE, is actually the culprit. The more common ApoE-3 is present in about 90% of the general population, but ApoE-4 is actually more common among people with AD. ApoE-4 binds to another protein in the brain, so that the two proteins together form clumps of protein throughout the brain. Scientists have found that the tendency to form these protein clumps is determined by the gene dose of ApoE-4. The average age at onset of AD is only 68 for someone who has two copies of the ApoE-4 gene, and is 76 for someone with one copy of the ApoE-4 gene, while someone who lacks the gene is unlikely to suffer dementia until about age 84. In fact, having two copies of the ApoE-4 gene is virtually a guarantee of AD before the age of 80.

A blood test for the ApoE-4 gene is now available, and careful analysis of a large number of family members of patients with AD has been done. We now know that having two copies of the ApoE-4 gene increases the risk of dementia more than eightfold with respect to someone who has no copies of the gene.[74] Even having a single copy of this gene increases the risk of AD nearly threefold with respect to normal. These conclusions are based on a large study, involving 235 people in 46 different families, so it is quite likely the data are correct. Confirmation has come from many quarters, as scientists all over the world have been able to replicate these results. One confirmatory study found that, among nearly 500 people thought to have nonfamilial AD, about 64% have at least one copy of the ApoE-4 gene, whereas in a normal population of people only 31% have one copy of this gene. Yet the fact remains that not all people with two copies of the ApoE-4 gene get AD. About 5% of people with two copies of the ApoE-4 gene do not have AD even near the age of 80; this suggests that there are other genetic sources of risk, and that inheritance of the ApoE-4 gene cannot explain all AD cases.

A recent study of 401 different families proved conclusively that there are multiple causes of AD.[34] A genetic analysis of over 2000 people showed conclusively that AD is heritable and not random, that there are a relatively small number of genes involved in producing the disease, and that women are more prone to AD than men. But it also became clear that more than one gene is involved, and that the early onset form of AD is distinctly different from the late-onset form. Among early onset AD families, the disease is caused by a dominant gene with a frequency of about 1.5% in the general population. Evidence was also obtained that a substantial fraction of late-onset AD cases are caused by new mutations, meaning that the disease is more complex than first thought, and that there may be many ways to produce a dementia that can be mistaken for AD. Having the ApoE-4 mutation does not cause AD outright, but it is a strong risk factor for the disease. Finally, it was shown that environment alone could be ruled out as a cause of AD in every case. While we may not understand everything about AD, it is very clearly a disease with genetic roots; the environment may be

contributory in some cases, but the environment is never sufficient to explain disease development.

Autism as a Heritable Condition

Autism is a very rare mental disorder that is discussed here
because it is strongly heritable and because it seems to provide a
bridge between overt mental illness and some of the more subtle
personality disorders which we will explore in the next chapter.
Autism is a strange disorder in which a child seems to lose
contact with the real world and is somehow insulated from
emotional contact with others, as if he or she existed in a glass
bubble. The classical symptoms of autism are fourfold: a preference for aloneness; an insistence on sameness; a liking for
elaborate routines and rituals; and the presence of some unusual
abilities that seem all the more remarkable because of the
concurrent presence of great deficits. Autism is fortunately quite
rare, as it afflicts about one in every 1000 children, but it affects
boys two to four times as often as girls. Autism appears to be
strongly linked to mental retardation: 56% of children with an IQ
measured at less than 50 were autistic, whereas less than 0.1% of
children with an IQ above 50 were autistic.[75]

Strangely, autism is also associated with very remarkable
mental abilities about 10% of the time. Musical ability can be
highly developed, as it was in a remarkable blind boy who
gained fame at the time of the Civil War. This child was thought
to be retarded and had never taken piano lessons, yet at age 11
he was able to play long classical compositions by memory after
hearing them only once. He was tested by a group of professional musicians, who played two piano compositions for him
for the first time, one a piece that was 13 pages long, the other a
20-page piece. He was able to play both perfectly from memory,
without hesitation and without seeming effort. Mathematical
ability can also be unusually well developed in the autist; an
autistic man in the 18th century was able mentally to calculate 2

raised to the 140th power, which is a 78-digit number. The calculation took several months, but it was apparently carried out unconsciously, as the man was able to conduct himself normally until the finished computation thrust itself into his conscious mind. Another autistic child was able to memorize the whole of Gibbon's *Decline and Fall of the Roman Empire,* and could recite it from memory, always making the same mistake, which seemed to reveal that he understood none of what he read. Oliver Sachs, the neurologist who wrote *The Man Who Mistook His Wife for a Hat,* has written about a severely autistic child who was able to draw so well and so fluidly, often completely from memory, that several volumes of his drawings have been published. Children with this uncanny sort of mental ability, despite apparent mental retardation, have been called *idiot savants.* Such remarkable mental abilities force us to question our definition of intelligence; if these gifts are not signs of intelligence, then perhaps intelligence is too narrowly defined.

Autism is not caused by any sort of psychological trauma in infancy, although there is some evidence that early physical injury of the brain can result in autism. But the evidence for a genetic basis for autism is stronger. Identical twins are more likely to be concordant for autism than are fraternal twins. The odds that a family with one autistic child will have another child so afflicted are 50 to 100 times greater than expected by chance alone.[75] Autism results from an impairment of communication and socialization, and all of the odd manifestations of the disease, including muteness, aloofness, repetitive play, obsession, and inappropriate social responses, may result from a root impairment of understanding. Some scientists believe that the separate threads of behavior in the autist result from the absence of a powerful integrating force, so common in most people that we are unaware of it; the search for meaning. An impaired ability to search for meaning must mean that some vital higher-order function of the brain is missing. That this higher-order function may be hereditary is very illuminating.

Table 1
Mental Illness in Identical and Fraternal Twins[a]

Relationship	Correlation	No. of twin pairs	Heritability
Autism			
Identical twins	60%	NA	90%
Fraternal twins	0%	NA	
Manic-depression			
Identical twins	62%	259	80%
Fraternal twins	18%	423	
Clinical depression			
Identical twins	77%	68	60%
Fraternal twins	45%	109	
Schizophrenia			
Identical twins	39%	310	50%
Fraternal twins	10%	480	
Attention deficit disorder			
Identical twins	24%	79	40%
Fraternal twins	0%	58	

[a]Data about each mental illness abstracted from sources as follows: autism[76]; manic-depression[54]; clinical depression[64]; schizophrenia[77]; and attention deficit–hyperactivity disorder.[78] NA indicates that the number is not available.

Genes and Mental Disorder

Given the striking reversals that have beset several high-profile genetic studies of mental disorder, it is understandable that many nonscientists are confused, discouraged, or even cynical about our chances of identifying the roots of mental disorder. Even though a role for genes is clearly shown by many split twin studies of mental illness (see Table 1), the actual identity of those genes remains elusive. And even in the strongest case, the association between the ApoE-4 gene and Alzheimer's disease, the gene is only thought to be responsible for 40–50% of all cases, which leaves a great many cases with unknown causes.

Yet we cannot lose sight of what progress has been made. We have witnessed a paradigm shift in the last two decades, a bloodless revolution of ideas in psychiatry. Psychiatry has changed from a practice to a science, and neuropsychiatric research is attracting some of the best young scientists. Mental illness is no longer always an intractable condition requiring long-term incarceration or endless therapy; most patients are now treated effectively with drugs having minimal side effects. We are beginning to understand mental processes at the cellular and subcellular level, a feat that would have been almost unthinkable a generation ago. Our understanding of mental illness lags behind our understanding of other mental processes such as learning, but the tools are in hand to eventually explain mental illness at the cellular and subcellular level.

Behavioral geneticists can now determine whether a particular gene is associated with a mental illness without any knowledge of what the gene actually does. And it will not be too much longer before knowledge of the structure or location of a gene will reveal key aspects of gene function. At that point, it will become possible to determine with surety the relative role of genes and environment in producing mental illness. This should eventually make it possible to block or reverse the progression of certain mental disorders before they become debilitating. With such a tantalizing goal in sight, it should not be surprising that many scientists continue to work toward a better understanding of mental illness.

10

Personality Traits

Much of what is colloquially meant by personality is the result of either intelligence or mental idiosyncrasies that, in their extreme form, become mental illness. Thus, it would seem that, in the preceding chapters on intelligence and mental illness, we have already discussed some aspects of personality. It is then a fair question to ask, what do psychologists or psychiatrists mean when they speak of personality?

Certainly, the general emotional state of a person is an important component of personality, but this is very hard to measure objectively. The state of relations between one person and the others around him are also critically important, and also hard to quantify. The factors that motivate an individual are central to their personality, but even people with a great deal of self-knowledge may be puzzled by some of their own actions, and it is not uncommon for people to have a poor grasp of their own underlying motives. Similarly, interests and aptitudes can define a person, but so many people share interests and aptitudes that this would seem an imprecise way of distinguishing between people. Finally, attitudes hold a central place in one's personality, but attitudes are often deeply held and highly emotional without having a strong rational basis. Consequently, people often have a difficult time reflecting on or discussing their own attitudes. And even a trained psychologist or psychiatrist may need lengthy sessions with a patient before the deeper motivations of that person can be understood.

For these reasons, scientists have developed a battery of tests that measure several different aspects of personality, each of which is thought to form a continuum and each of which interacts with the others. Psychologists assess personality by having a subject answer a wide range of questions that relate back to these key aspects of personality. Questions are usually multiple-choice and cover a broad range of material, so as to take a broad inventory of personality traits. Multiple-choice questions are preferred to oral or essay tests for several reasons: they permit a broad range of coverage, they reduce the element of chance in the choice of questions, they place all test-takers in an equally uncomfortable situation, and they eliminate the possibility of favoritism for some test-takers. While these tests may miss the more subtle aspects of personality, they do at least provide an objective descriptor of those ineffable traits collectively known as personality.

Yet it is critical to note that there is a fundamental difference between tests of personality and tests of intelligence.[79] Personality is measured subjectively, by self-report or self-rating of behavior, whereas intelligence is measured objectively, by solving difficult test problems. If intelligence was measured in the same way as personality, each person might be asked to rate his own vocabulary or mathematical ability, then these self-ratings would be assumed to be objective and correct. Though it sounds silly, this is exactly what those who take a personality test are asked to do.

What Is Personality?

Personality is thought to be comprised of five key determinants, each of which spans a broad range, and each of which can vary independently of the other determinants. An individual expresses each of the determinants to a greater or lesser extent, forming a unique personality from "building blocks" that are common to all people. Because there are only five such building blocks, the system assumes that all people can be fully categorized on the basis of these determinants only. In essence, this amounts to an unappealing "one from column A, one from column B" approach to describing personality; it seems that personality is really far too complex to be adequately described by such a simple system. In any case, the separate determinants, together with an indication of the range each determinant can span, are as follows[25]:

High level of:	Low level of:
Extroversion:	
(social dominance, self-involvement, emotionality)	
Outgoing, decisive, persuasive, enjoys leadership roles	Shy, retiring, reserved, withdrawn, does not enjoy being the center of attention
Neuroticism:	
(anxiety, emotional stability, reactivity to stress)	
Emotionally unstable, nervous, irritable, prone to worry	Emotionally stable, quickly recovers from upsetting experiences, not prone to worry
Conscientiousness:	
(conformity, dependability, authoritarianism)	
Organized, responsible, practical, dependable, good planner	Impulsive, careless, irresponsible, cannot be depended on
Agreeableness:	
(likability, friendliness, aggression)	
Sympathetic, warm, kind, good-natured, does not take advantage of others	Quarrelsome, aggressive, unfriendly, cold, vindictive
Openness:	
(culture, intellect, imagination, sophistication)	
Broad interests, insightful, curious, original, imaginative, open to novel experiences	Narrowly defined interests, unintelligent, unreflective, shallow

There is consensus, although by no means unanimity, that these five determinants comprise the essence of personality. In principle, each of the determinants can vary independently of the others; it should be possible, for example, to combine a high level of extroversion with a low level of agreeableness in the same person. But this seems far less likely than a person who combines a high level of both extroversion and agreeableness. In practice, the separate determinants may not be so separate after all, making this system of describing personality somewhat suspect.

There also seem to be important dimensions of personality that are not adequately brought into this system. For example, mood can dramatically alter personality (as anyone who has interacted with a manic-depressive would know), yet mood is not measured in any way. And different traits could interact with each other in subtle ways; a person might normally be somewhat introverted but also impulsive, so that they might appear to be extroverted in certain social situations. Impulsivity would not necessarily make a person any less conscientious or less responsible, but it might act to overcome a natural tendency to shyness at certain times. It is also difficult to imagine how a mental problem as common as a phobia would fit into this system. It seems plausible that someone could be emotionally stable, a trait that is usually associated with a low level of neuroticism, yet at the same time be badly claustrophobic; under the wrong circumstances this person would appear to be highly neurotic.

This simple system of classifying human personality may thus fail to capture the full range of possible personality traits. It also seems likely that this system would fail to differentiate between two people with similar, yet not identical, personalities (e.g., fraternal or identical twins). Some scientists would like to use a more complex system of personality classification, one that incorporates more determinants, in an effort to capture subtler variations in personality. Yet at some point this additional complexity might make the system too unwieldy to use. It is also quite likely that, because the circumstances under which a test is given are rigidly standardized, the test is accurate insofar as all test-takers share the exact same surroundings during the test. In other words, it would not matter that one person is claustrophobic while another is not, because the test is not administered to the claustrophobic person while he is kept in a small room. Scientists thus routinely compromise on the five-determinant system because it is practical, easy to administer, and already widely used. No matter what system eventually becomes established, this simple system is at least adequate for discussing the heritability of personality.

How Is Personality Measured?

Personality testing began as an offshoot of the effort to measure intelligence, but the effort to measure intelligence is really quite ancient. Three thousand years ago, in 1115 B.C., the Chinese government began to require applicants to the civil service to undergo tests to determine who would be best able to handle the demands of the job. The first time that intelligence tests were very widely used in the United States was in 1917, when the United States entered the First World War. A committee was appointed by the American Psychological Association to determine ways in which psychologists could assist in the conduct of the war. It was recognized that a method was needed to rapidly classify one and a half million new Army recruits, with respect to general intellectual level. This information could be used to determine who was eligible for service, what tasks the eligible entrant should be trained for, and who among the entrants was best qualified for officer training. For the first time, intelligence level was tested with a multiple-choice examination that could be administered to many entrants simultaneously. The test was designed so that instructions to test-takers were quite simple, and no special training was needed to administer the tests. After the war, these tests were released to the public by the Army, and schoolteachers soon began to administer the same tests to students.

The prototype of the personality test, or self-report inventory, was also developed during World War I, as a way to identify seriously disturbed men who were unfit for service. The test had a number of questions that addressed common symptoms of mental illness, and it was scored by counting the number of such symptoms reported. Unfortunately, the personality test was not finished early enough in the war to be widely used, but it was released afterwards to the general public by the Army. Release of the personality test occurred at a time when psychologists were interested in formulating generalized descriptions of human behavior. The uniformities among people, rather than the differences between them, were thus of greatest interest. There was special interest in characterizing traits that could be broadly explanatory of human behavior, so personality testing was soon done on large numbers of people. Personality testing has nonetheless lagged behind intelligence testing, in terms of accomplishments, because personality is such a difficult thing to describe: subtle, yet central to the well-being of the individual; complex, yet often fairly stereotyped in day-to-day interactions; and rarely ever fully explored, even by the individual in question.

Modern personality testing uses three different approaches, sometimes in concert, to gain an understanding of the individual. The modern self-report inventory is similar in essence to the test developed during World War I, and relies on a subject voluntarily revealing information about traits that are characteristic of mental imbalance of one sort or another. Situational or performance tests involve the performance of a structured task, with the real object of the exercise being hidden from the test-taker. The test is often designed to reveal behaviors such as lying, cheating, stealing, or persistence. This kind of testing is less common nowadays because it requires specialized test facilities, trained personnel to do the testing, and it is somewhat subjectively scored. The final type of personality test are the projective tests, which are quite popular now. These tests involve completion of a relatively unstructured task that has a wide range of possible solutions, on the assumption that personality will play a major role in completion of the task. Projective tests include such things as free association of words, sentence completion, drawing, interpretation of the famous Rorschach ink blots, or role-playing in situations drawn from real life. For each of these three broad approaches, there are a number of different types of test available.

Broad consensus has emerged that personality is a combination of temperament and character. Temperament is essentially the inborn or genetic predisposition to an emotional state, whereas character is the evolving element of one's personality, forged in the fire of life events both within and without the family. Clearly, temperament has an impact on what character will evolve, perhaps because temperament can influence the type of environment an individual selects for himself. But, just as clearly, environment and life events have a major impact on personality. Thus, the goal of personality testing is to character-ize the products that result from a combination of temperament and character, including emotional state, interpersonal relation-ships, motivations, interests, and attitudes.

Differences in Temperament of Infants

Evidence for the inheritance of temperament is fairly strong, and temperament can manifest itself at a very young age.[80] Components of the personality can become apparent at a few months of age and can remain strikingly constant thereafter, during the process of maturation. For example, infants only 4 months of age were examined, to determine their reaction to visual stimulation and their characteristic level of motor activity. In a group of 600 white infants, roughly 35% were characterized by infrequent crying and low motor activity and were thus classified as having low reactivity, while another 20% were characterized by frequent crying and high motor activity and were thus classified as having high reactivity. Of the remaining infants, 25% were found to cry frequently but have low levels of motor activity, while 10% cried infrequently but had high levels of motor activity. These same subjects were evaluated again at 14 months of age, to determine which infants were fearful and which were most at ease in unfamiliar or threatening situations. Of those children with low reactivity at 4 months of age, the overwhelm-ing majority (62%) were not fearful at 14 months of age. Con-versely, of those who were highly reactive at 4 months of age, the overwhelming majority (90%) were moderately to highly fearful at 14 months of age. These same trends were also present at 21 months of age, when the subjects were tested again. Observation of the interaction between mother and child showed that the mother's behavior had relatively little impact on whether or not the child was fearful. Evidently, reactivity or fearfulness is a component of the personality that is set at only a few months of age, and is remarkably consistent through the first 2 years of growth. Scientists believe that fearfulness is a consequence of the general state of arousal of the child, so that a chronically aroused child is much more likely to be fearful. There is also evidence, from research with animals, suggesting that the level of chronic arousal is actually hardwired into the brain.

This same examination, which was initially given to white babies in Boston, was also given to babies of the same age in Beijing, to determine whether any consistent differences would emerge between these two groups.[80] Under identical conditions, American babies spent about fourfold more time moving around than did Chinese babies, and were fivefold more likely to fret or cry. Chinese babies were generally less irritable than American babies, but were also less likely to vocalize. However, both groups of babies spent about equal amounts of time smiling at the examiner. These findings suggest that there are characteristic differences between Chinese and American infants, in terms of the degree of arousal. By the criteria used in this test, American infants appear to be chronically aroused to a greater extent than Chinese infants. Interestingly, adult white Americans with symptoms of anxiety often require higher doses of sedative than do adult Asian-Americans with the same symptoms.

The Genetic Basis of Personality

As expected, the clearest insight into the heritable nature of personality has come from split twin experiments. A study that provided recent insight examined 56 sets of identical twins who had been separated at birth and reared apart from one another.[8] Each twin was put through a battery of 50 hours of medical and psychological assessment, so that a very detailed portrait of each twin could be developed. Because these particular twins shared all of their genes but none of their environment, similarities between twins could arise only from the genes, whereas differences must be attributed to the environment. Not surprisingly, there were a great many physical similarities between the twins. Twins were 97% identical in the pattern of their fingerprints, indicating that fingerprint patterns are about 97% heritable. Similarly, the heritability of height was 86%, while that of body weight was 73%. Heritability of blood pressure was 64%, while heart rate, which is very strongly influenced by physical conditioning, was still 49% heritable. The heritability of physical attributes in these twins provides a very useful context within which to consider the heritability of personality traits.

Personality variables were measured in these identical twins reared apart, and personality was about as heritable as heart rate overall.[8] In total, 11 different subtests on the Multidimensional Personality Questionnaire were given to 44 sets of identical twins, and another 18 different subtests of the California Psychological Inventory were given to 38 sets of twins; this is an impressive amount of data from which to generalize. Overall, about 50% of all personality traits that could be assessed by testing were shared between identical twins, even though these twins shared no environment. These shared traits must be the result of shared genes, which is astonishing evidence for the heritability of personality. In fact, identical twins reared together were no more similar for personality traits than were identical twins reared apart, which is convincing testimony to the power of the genes in determining personality. That identical twins reared together are really no more similar to each other than identical twins reared apart may even indicate that the actual "shared environment" of twins is not all that important in molding their personality. It is perhaps reasonable to speculate that genes affect personality indirectly, by influencing how an individual structures or experiences his own environment. In other words, a functionally similar environment might be created for identical twins, whether or not they are actually reared together.

In evaluating the similarities between identical twins reared apart, it is important to bear in mind the strengths and weaknesses of the personality tests themselves.[8] If the same person is tested twice, using a Multidimensional Personality Questionnaire, the similarity from one occasion to the next is about 88%, meaning that the test is reasonably precise. The fact that an individual tested twice does not reproduce his own answers exactly may be related to the fact that personality is actually fairly changeable in a person, as that person ages. Thus, the measured similarity between identical twins in this study may arise, in part, because sets of twins were tested at the same chronological age. Unrelated sets of people tested for personality will share only random commonalities, and these random commonalities are expected to be very small in a large sample of people. If data from unrelated sets of people were used to calculate heritability, it is likely that the calculated "heritability" would be a few percent at most, and that this number would approach zero with a large enough sample size. While our current personality tests undoubtedly have problems, the fact that identical twins reared apart are so similar is strong evidence for the heritability of personality.

Exhaustive testing of occupational, vocational, and general interests has shown that these facets of personality are about 40% heritable overall.[8] There were significant differences between identical twins reared together and identical twins reared apart, which confirms that the role of the environment is fairly large in determining interests. Finally, several different tests showed that even something as nebulous as social attitudes are about 40% heritable. In fact, the degree to which twins reported that religion was important in their life was 49% heritable, while the extent to which twins adhered to the same traditional values was 53% heritable. The biggest single difference that could be found between the twins was in "nonreligious social attitudes," which are apparently much influenced by education, but which were still 34% heritable.[8] While one should bear in mind that identical twins reared apart provide an estimate of the maximum plausible heritability of a trait, these results are still astounding. Parenthetically, IQ was found to be 69% heritable, on the basis of these identical twins reared apart (as we discussed in Chapter 8).

Table 1
Inheritance of Personality Traits[a]

Trait	Heritability	Environment
Extroversion	47%	53%
Openness	46%	54%
Neuroticism	46%	54%
Conscientiousness	40%	60%
Agreeableness	39%	61%
Overall personality	45%	55%

[a]Data for each personality trait averaged from many separate studies.[25]

Different personality determinants could conceivably differ in their degree of heritability, with some determinants being more heritable than others. However, current data do not show strong differences between determinants. Several years ago, a compilation of four different studies, which together involved over 30,000 pairs of twins, suggested that extroversion and neuroticism are both about 50% heritable.[22] More recently, data compiled from different studies (Table 1) show a fairly uniform heritability of the different personality traits; extroversion is most heritable at about 47%, and agreeableness is least heritable at about 39%.[25] These conclusions are based on several different studies that examined large numbers of twins, used modern methods of psychological assessment, and employed the very latest computer models to calculate heritability. Men and women differ somewhat, in that the heritability of personality is marginally lower in men than in women. But all studies converge in concluding that the heritability of personality traits averages between 41 and 51%, and that environment plays a substantial role in determining personality.

The most surprising aspect of the heritability of personality is that "shared environment" seems to account for a very small part of the "shared personality" of siblings.[27] Identical twins reared apart are typically about as much alike as are identical twins reared together, for traits like emotional stability. Furthermore, among genetically unrelated children, sharing a family

environment seems to produce no similarity at all in emotional
stability. These findings together suggest that shared environment plays a very minimal role in development of this personality trait at least. This means that the environmental factors
which psychologists have spent so much time agonizing over
may not really be critically important. If the current models of
personality inheritance are correct, such things as childrearing
practices, birth order of siblings, parental education, neighborhood, schooling, and television viewing are, in the aggregate, no
more important than genes. This will be a very contentious
conclusion for people who are imbued with the current paradigm in psychology, which says that family interactions are
central in forming personality. Yet there may be an explanation
that resolves this problem.

According to current models of the inheritance of personality, both "errors in measurement" and the "unshared environment" are more important than the "shared environment" in
accounting for personality differences between twins.[25] Shared
environment seems to account for about 7% of the shared
personality of twins, whereas error and unshared environment
each account for roughly 25% of the variation in personality.
Errors in measurement are a major problem because, in comparing twins, two separate personality tests must be given, and
each of these tests is likely to be somewhat in error. "Unshared
environment" includes all of those aspects of the environment or
experience of twins that are unique to one twin; this could
include differences in friends, or teachers, or parental interactions, or something else entirely. Such subtle unshared elements
of the environment appear to cumulatively account for fairly
major personality differences between one twin and the other.
Overall, studies suggest that "shared genes" are more than five
times as important as "shared environment" in determining
personality traits. Furthermore, "unshared environment" is
more than three times as important as "shared environment" in
forming personality.

To put this another way, temperament seems to induce a
person to structure their environment in a certain way, so that

unshared environment may bring out traits that are latent in the personality. Unshared environment could thus amplify the effect of genes on the personality, although environment probably cannot induce a new trait to be expressed in the personality without there being some genetic basis for that trait. In essence, unshared environment can conspire with genes to evoke a personality trait that, under other circumstances, might remain dormant. A child born with an antisocial attitude will almost certainly elicit less social interaction than normal from teachers and classmates, so that the antisocial attitude is, in some sense, validated for the child. Yet, if this child were to be placed in an infinitely "friendly" environment, the child might still be unable to respond because of his inherent antisociality. Alternatively, an extroverted and friendly child will evoke friendliness from those around him, which would likely reinforce the initial tendency to friendliness. The idea that genes can affect the personality indirectly, by influencing how an individual structures his environment, is conceptually appealing. Nevertheless, this idea cannot be accepted without considerably more proof than it now enjoys.

There can be no doubt that genes have a powerful influence on personality traits. Biological children of parents who have some exaggerated trait are likely to have the same trait. Genes have a powerful influence on the environment of a child, both through the structure that parents impose on the environment and through the structure that the child selects from the environment. Genetic effects on personality could even masquerade as environmental effects. For example, parents of children who are aggressive tend to use more punitive discipline, perhaps as a result of genes that confer both greater physical aggression on the part of the children and greater response to aggression on the part of the parents.[81] That genes can subtly influence the environment calls into question much of the research done on "socialization" of children, since most such studies have blithely ignored the possibility that genes can act so indirectly. Conversely, those influences on a child that are environmental can act through many routes unrelated to the family, including friends, teachers, employers, and other nonparental adults.

Good evidence seems to show that the heritability of personality traits declines with age.[22] This is consistent with the interpretation that we can indeed learn and grow as we age, and that some of this learning can modify personality traits. Yet this conclusion is at odds with evidence suggesting that the heritability of IQ increases with age. Even though IQ and personality are two different things, intelligence and personality are linked; it seems implausible that the heritability of IQ could increase with age, while the heritability of personality declines. Intelligence seems to determine, at least in part, the range of expression of personality traits. Low intelligence is often associated with troublesome behavioral problems and maladaptive personality traits, which again suggests a role for learning in the expression of personality traits.

In any case, genes do not absolutely decree personality; education, experience, or enlightenment can cause convulsive transformations of personality. One of the clearest examples of this is the story of a man, born of unreasoning hatred, who ultimately became that which he hated. This man was the son of a Ku Klux Klan leader and was raised within the folds of a Klan robe as an ardent anti-Semite. He spent much of his life circulating hate literature, burning crosses, and actively oppressing Jews. But, improbable as it seems, he formed a personal relationship with a Jewish couple whom he had persecuted, and he eventually converted to Judaism. Thus, for this man at least, neither genes nor environment entirely precluded the attainment of a whole and balanced personality.

Inheritance of Personality Disorders

Personality disorders are estimated to affect at least 10% of the population in the United States, so they are a very common problem.[82] Personality disorders can be quite disabling, and they may or may not occur in conjunction with either physical disorders or other psychiatric disorders. According to the American

Psychiatric Association, personality disorders are defined by the following elements:

1. An enduring pattern of inner experience and outward behavior that is markedly different from the cultural norm. Differences could exist in ways of perceiving and interpreting events and people, or in the range, intensity, and appropriateness of emotional response, or in relationships with others, or in the degree of control of impulsive behavior.
2. This aberrant pattern of behavior is inflexible and pervades the personality, creating significant distress or impairment for the individual.
3. The pattern is stable and of long duration, usually originating in adolescence, and persisting into adulthood without remission or exacerbation.
4. The pattern cannot be attributed to drug abuse, disease, trauma, or any recognizable mental illness.

As clear-cut as these elements seem on paper, applying them can be problematic. It is often difficult to discern between a personality disorder and what is simply a "difficult" personality, in the same sense that it is difficult to delineate the exact point at which normal blood pressure becomes high blood pressure. There is now a groundswell of support for the idea of defining personality disorders with objective criteria, using scores achieved on tests of the five personality determinants. There are presently ten known personality disorders (PDs), which may overlap each other somewhat, but which cluster into three groups, as follows[82]:

Cluster A—Odd/Eccentric

1. Paranoid PD—distrust and suspicion of others, so that the motives of others are always interpreted as malevolent
2. Schizoid PD—detachment from social relationships and a restricted range of emotional expression with others

3. Schizotypal PD—acute discomfort with and reduced capacity for close relationships with others, as well as cognitive and perceptual distortions and personal eccentricities.

Cluster B—Dramatic/Emotional/Erratic

1. Antisocial PD—disregard for and violation of the rights of others occurring since age 15
2. Borderline PD—instability of interpersonal relationships, fluctuating self-image and emotional state, and a strong component of impulsive behavior
3. Histrionic PD—excessive emotionality and attention-seeking behavior
4. Narcissistic PD—grandiose behavior and fantasies, strong need for admiration, and a lack of empathy

Cluster C—Anxious/Fearful

1. Avoidant PD—social inhibition, feelings of inadequacy, and hypersensitivity to criticism
2. Dependent PD—excessive need to be taken care of that leads to submissive behavior and fear of separation
3. Obsessive-Compulsive PD—preoccupation with ritual, order, perfection, and mental and interpersonal control, at the expense of flexibility, openness, and efficiency

Clearly, some of these PDs are similar to personality traits shared by many people, so it is more a question of degree than kind. It is also clear that different PDs have different degrees of heritability, but that all apparently result from an interaction between genes and the environment. Both schizoid PD and schizotypal PD may identify someone at risk of developing schizophrenia, and both tend to have a fairly large hereditary component.[82] Similarly, paranoid PD appears also to be strongly hereditary and may be linked to paranoid schizophrenia. But the environment also plays a role, as both trauma and childhood abuse seem to contribute to development of borderline PD, while an excessively rigid and moralistic family environment may contribute to development of obsessive-compulsive PD. Generally, it seems that a nurturing family environment can protect a child from developing PDs, and can at least partially offset a genetic vulnerability to PD.

Personality disorders are a major societal problem whose magnitude has been persistently underrated.[82] They are associated with unemployment, underemployment, job inefficiency, lack of job satisfaction, disability, and substance abuse. Antisocial PD, for example, can result in a loss of income comparable to that resulting from substance abuse, and in excess of the loss of income associated with schizophrenia. Individuals with antisocial PD are at a higher risk than normal of suicide, homicide, and mortality from accidents. In general, persons with PDs are more than twice as likely as normal to be hospitalized at some point in their life.

The Origins of Personality Disorder

The development of new computer models of inheritance has made it possible to gain an insight into the origins of PD, by analyzing data gathered from identical and fraternal twins *not* separated at birth.[83] Even though such twins grow up together, and thus share both genes and environment, these factors can still be separated mathematically. Yet this is possible only because of knowledge garnered from past studies of twins who were separated physically at birth. One recent study measured the similarity of personality test scores for 90 identical twins and 85 fraternal twins, using a test called the Dimensional Assessment of Personality Pathology (Table 2). This test is quite involved, since it requires each individual to answer 290 questions, which together characterize 18 different pathological personality types; these pathological personalities are related to the PDs described above. Several interesting findings emerged from this work. First, it was found that males had consistently higher levels of certain types of personality pathology than did females

Table 2
Twin Similarities for Personality Pathology[a]

Personality pathology (brief description)	Twin correlation units shown below		Estimate of heritability
	Identical	Fraternal	
Distorted perceptions (stress psychosis)	82%	39%	41%
Narcissism (attention-seeking, need for approval)	64	12	64
Callousness (lack of empathy, sadism)	63	29	56
Identity problems (self-contempt, pessimism)	60	26	59
Stimulus-seeking (reckless, impulsive)	59	33	50
Passivity (lack of organization)	58	27	55
Social avoidance (poor social skills, shyness)	57	26	57
Restricted emotions (self-contained)	51	25	47
Rejection (judgmental, rigid, dominating)	49	22	45
Suspiciousness (suspicion of others' motives)	49	25	48
Emotional instability (anger, irritability)	48	13	49
Intimacy problems (inhibited, low sexuality)	40	0	38

[a]The similarity between identical and fraternal twins is shown for elements of personality pathology; each listed trait is significantly more often shared by identical than fraternal twins.[83] Similarity is indicated by the value of the twin correlation; a 100% correlation would mean that twins are exactly alike, whereas a 0% correlation indicates no shared similarities at all.

(in particular, males had higher scores for callousness, narcissism, rejection, restricted emotions, stimulus-seeking, behavior problems, and suspiciousness). Second, age was related to many of these traits, in that older twins were less likely to suffer from these pathologies of personality than were younger twins (i.e., older twins tended to have lower scores for anxiety, emotional instability, distorted perception, callousness, narcissism, behavior problems, passivity, stimulus-seeking, and self-harm). Finally, and perhaps least surprisingly, identical twins tended to be much more similar to each other than were fraternal twins. This strongly suggests that all of these personality pathologies have their origin, at least in part, in the genes.[83]

According to the computer model, genes alone were responsible for about 44% of all of the personality pathologies identified. Although this number has a large margin for error, given that it is estimated using a complex computer model, it is in line with prior estimates of the impact of genes on personality. Two-thirds of the 18 personality pathologies examined had an estimated heritability between 40 and 60%, consistent with estimates of the heritability of normal personality traits. This is interesting because it implies that PDs are simply extremes of the normal range of variation in a personality trait. The personality pathology most strongly related to genes was narcissism, the need for constant adulation and approval, which often motivates attention-seeking behavior and a grandiose manner; narcissism was estimated to be about 64% heritable. Perhaps surprisingly, identity problems, which include chronic feelings of emptiness, self-contempt, and pessimism, were calculated to be 59% heritable. Social avoidance, which is social apprehensiveness or shyness, combined with a low need for personal affiliation, a fear of interpersonal hurt, and often, defective social skills, was thought to be about 57% heritable.

Overall, the environment was responsible for explaining an average of 56% of all personality pathologies in twins. Environment played a relatively minor role in explaining narcissism, identity problems, social avoidance, callousness, and passivity.

The weak role of the environment in creating these maladies may mean that they would be difficult to treat with counseling or psychoanalysis alone. On the other hand, environment played a major role in creating insecurity, intimacy problems, submissiveness, compulsiveness, and self-injurious behavior. In fact, environment was thought to explain all of the conduct problems seen in twins, including juvenile antisocial behavior, interpersonal violence, addictive behavior, and failure to adopt social norms. Therefore, these conditions might be considerably more responsive to therapy, since the genes play a lesser role in expression of these personality pathologies.[83]

Inheritance of Learning Disorders

Although learning disorders are quite distinct from PDs, learning disorders will be discussed here because they are often first manifested as a personality problem that disrupts the classroom. The most familiar learning disorder is attention deficit disorder (ADD), which occurs with or without hyperactivity. The child with ADD is restless, inattentive, impulsive, and easily distracted, so this disorder can cause learning problems for the child and discipline problems for the entire classroom. As discussed in the last chapter, ADD is related to manic-depression, and it is known to be strongly familial.[56] One study found that more than a third of children with ADD had some other type of emotional disorder as well, with clinical depression and manic-depression being the most common. This is a striking finding because the current clinical picture of ADD does not include symptoms of mood disturbance. Relatives of children with ADD were also at a fivefold higher risk of manic-depression than normal. These results suggest that there can be a genetic predisposition to both ADD and manic-depression, and some scientists have even suggested that ADD may actually be an early manifestation of an emotional disorder.

There is also good evidence for a genetic cause of the learning disorder known as dyslexia.[84] Dyslexia is a marked deficit in reading ability which affects 5–10% of school-age children, even though these children have average intelligence and adequate educational opportunity. Dyslexia is two- to three-fold more common in boys than girls, and is associated with a slight reduction in IQ, although it is possible that the measured reduction in IQ is related to the fact that most IQ tests require a certain level of reading proficiency. The first strong indication of a genetic basis for dyslexia came from a study of 119 twin pairs, at least one of whom had reading disability. A total of 64 pairs of identical twins and 55 pairs of fraternal twins were studied, all about the age of 13. Each child identified with reading disability was extensively tested, to verify that the reading disability was not related simply to poor schooling, and to determine whether the child was of normal intelligence. Children were excluded from the study if their IQ was below 90, or if they showed any evidence of behavioral or emotional problems, or if their vision or hearing was worse than normal. As expected, the degree of concordance for dyslexia was higher among identical twins than among fraternal twins. In fact, the degree of heritability was estimated to be about 60%, meaning that roughly two-thirds of the reading deficit is related to genetic factors that are completely independent of the environment.

More recently, dyslexia has been attributed to a gene that is found in a small region of human chromosome 6.[85] This association was first identified because children with dyslexia are prone to develop autoimmune diseases which also are associated with genes that map to chromosome 6. The children evaluated in this study were even more stringently screened than in the previous study; children had to have an IQ above 90 and be free of other problems, but they also had to come from a family with several generations of reading problems and they had to be delayed in their reading by at least 2 years. Given the stringency of these requirements, it is not clear that the findings of this study will be broadly applicable. In any case, it was found that a gene related to dyslexia maps to chromosome 6, and there is only a 1 in 100,000 chance that this gene was erroneously identified. Thus, a gene on chromosome 6 may be responsible for severe dyslexia. Yet this does not mean that other dyslexia genes do not exist; in fact, it is possible that several other genes could contribute to mild cases of dyslexia. But the major dyslexia gene on chromosome 6 accounts for roughly 50–70% of the heritable variation in this learning disorder.

This work provides a ray of hope because, if dyslexia is indeed heritable, then determining the genetic basis could provide insight into possible causes, could allow prospective parents to have an accurate indicator of whether their children will be affected, would facilitate earlier diagnosis of the problem, and might even suggest possible treatments. Yet information about genetic tendencies can also be misused. For example, a child diagnosed at an early age with a reading disability might be directed into an academic track where little is expected of them, and so little is given to them, in the way of educational resources. In the worst case scenario, it is possible that genetic information could even be used to screen a fetus for possible abortion. Advances in molecular genetics are rapidly making human engineering possible, while the eugenic practices of this century have already shown that we have the ability to pervert ideals and totally ignore ethical considerations.

Genes and Personality

Evidence for a heritable basis of personality is currently weaker than the evidence for a heritable basis for either intelligence or mental illness. This is because both intelligence and mental illness are easier to define, easier to measure objectively, and hence easier to study than is personality. It seems that the present system of defining personality, on the basis of only five key determinants, is perhaps too crude to capture the nuances of personality shown by most people. This would mean that our current understanding of personality is probably inadequate. Yet this does not argue against either studying personality or trying to generalize from what we do know.

Whether or not current testing procedures are adequate to understand the more subtle aspects of personality, it is clear that those aspects we can now measure are often shared by members of a family. Studies of identical twins reared apart have shown that personality variables are roughly half inherited and half environmentally mediated. Strikingly, identical twins reared together are no more similar in personality than are identical twins reared apart, which clearly suggests that genes find a way to assert themselves. How genes actually work to mold personality is not understood, yet it should be remembered that personality is no greater than the sum of its parts; just as learning and memory will eventually be explained at the cellular or subcellular level, so too will personality eventually be understood as an outcome of the dialogue between nerve cells. In this admittedly reductionist viewpoint, personality is seen as a kind of weighted average of all possible interactions between nerve cells. Some of these interactions will be more likely than others for an individual person, which might explain one person's proclivity to extroversion or another person's tendency to shyness. In other words, what causes a person to be extroverted rather than shy may be a kind of molecular bias built into the synapses between certain nerve cells.

This view makes it somewhat easier to imagine how genes could play a role in personality, since they could act by building in the molecular bias in the first place. This is all very speculative, of course. We certainly cannot yet explain personality at the cellular level, and it may be that personality will never be fully explained at this level. But we must not conceive of personality as something too ineffable, too complex, or too sacred to study, or we will never learn anything more about it. Acknowledging the role of the genes in helping to create personality is an important step toward gaining a more complete understanding of the underpinnings of our own humanity.

11

Sexual Orientation

Surely one of the most controversial areas of human behavior is that of sexuality and sexual orientation. Both psychology and behavioral genetics are now wracked with divisive arguments about the origins of sexual orientation. In fact, psychologists have been arguing about what causes homosexuality for more than 100 years, and they are actually further from consensus now than they were a decade or two ago. Psychological explanations of homosexuality, based on the teachings of Freud, were widely accepted until fairly recently. But Freudian explanations have wilted in the face of conflicting data, and most psychologists have concluded that a new theory of sexual orientation is needed. Similarly, behavioral geneticists are deeply divided about the validity and significance of several recent high-profile studies on the inheritance of sexual orientation.

Freud believed that every child is inherently bisexual at birth, but during a particular stage of development that occurs at about 4 years of age, this changes. Children learn to suppress sexual feelings toward members of the same sex, and to direct sexual feelings toward members of the opposite sex. Freud proposed that male homosexuality originates when this developmental stage is blocked, perhaps because of the presence of a domineering mother figure or the absence of a strong father figure. This "psychodynamic theory" of homosexuality predicts that homosexual men would tend to be emotionally distant from their father and perhaps emotionally dependent on their mother. However, when this prediction is actually tested, it is usually found to be quite weak.

Toward a Biological Explanation of Homosexuality

Beginning in the 1950s, animal research began to show a very potent effect of hormones on the development and differentiation of the brain. It was found, for example, that the sexual behavior of rats and mice could be radically altered by exposure to the male hormone testosterone during a critical window of time during development. Of perhaps more relevance to humans is the fact that female monkeys exposed to testosterone during prenatal development tend to play with other monkeys in a manner more typical of male monkeys. This latter observation is especially intriguing because some scientists contend that, in childhood, boys who will later become homosexual tend to play in a manner more typical of girls. Early play may thus be revealing the incipient presence of a male or female identity, shaped in part by hormonal exposure prior to birth. But, if it is true that prenatal hormone exposure can affect later sexual orientation, one must wonder what the factors are that determine prenatal hormone exposure. It is of course possible that prenatal hormone exposure is affected by both genetic and environmental factors, since stress can affect hormone secretion.[86]

Early observations on the link between hormone exposure and animal behavior led to development of a "neurohormonal theory" of sexual orientation. This theory has received much support recently from a wide range of different observations and experiments.[85] For example, it has been noted that women who were prenatally exposed to DES (diethylstilbestrol, a potent blocker of the female hormone estrogen) tend to report homosexual feelings more often than women who were not prenatally exposed to DES. Since DES was extensively used to maintain problem pregnancies during the 1950s, a substantial number of women were prenatally exposed to this artificial hormone. The fact that women exposed to DES tend to have sexual feelings more typical of men suggests that this hormone can have an effect on humans that is at least conceptually similar to the effect of testosterone exposure in monkeys. It is also known that early childhood play typical of the opposite sex is the single best predictor of homosexuality, which implies that sexual orientation must be set at a very young age, or perhaps even prior to birth.[87]

But far more convincing evidence of a hormonal basis for homosexuality has been obtained recently. Several very striking hormonal anomalies have been discovered in gay men, suggesting that homosexual men may have hormonal responses that are intermediate between heterosexual men and heterosexual women.[88] In one study, homosexual and heterosexual volunteers were recruited by advertisement and by word of mouth, and volunteers were subjected to an in-depth interview to establish their sexual orientation. Then a hormone was administered to a group of 14 avowed gay men, and the physiological responses of these men were compared to those of 17 straight men and 12 straight women. A single dose of estrogen was given to individuals in all three groups, and then the amount of a second hormone, called luteinizing hormone (LH), was measured in the bloodstream. In straight women, estrogen administration caused a dramatic rise in LH over the course of 3 days, while in straight men estrogen actually caused a depression of LH over the same time period. In gay men, estrogen caused a rise in LH similar to, and almost as large as, the rise seen in straight women. These results suggest that gay men have hormonal responses that are midway between that which is typically male and that which is typically female. There are any number of reasons why this is so, but the important point is that there seems to be an actual biological indicator or "marker" of homosexuality in at least some gay men.

This study, like most other studies, had faults that weaken the conclusions, but overall the study was fairly strong. It is possible that gay men willing to participate in such a study are not typical of all gay men; many gays are not comfortable enough with their sexuality to participate in a study that requires the open admission of sexual orientation. In fact, men in this study were chosen so that they may have represented opposite ends of a spectrum of sexual preference. Heterosexual men in this study reported having had no homosexual experiences since puberty, and the homosexual men reported a pattern of male-oriented sexual behavior since puberty. According to Kinsey, only 4% of gay men are exclusively homosexual after puberty, so this may have been an unusual group of gay men. Several recent studies have been unable to completely replicate this early study, so it is also possible that the results are simply wrong. Nevertheless, the early study found several fascinating differences between gay and straight men, and these differences are consistent with a hormonal theory of homosexuality.

It should be noted that the gay community strongly favors a biological explanation, rather than a psychodynamic explanation, for homosexuality. While it is possible that gays favor a biological explanation because they have greater insight into the origins of their own sexuality, there may also be a hidden agenda at work. The psychodynamic explanation of homosexuality implies an element of personal choice in being gay, whereas the hormonal explanation of homosexuality implies no such choice. If no element of choice is involved, then homosexuality is more like a fact of life or a biological necessity, than a moral failing or perversion. Many gays believe that tolerance from the straight community is more likely to be achieved if straights perceive that gays had no choice but to become homosexual. Thus, there could be a reason for at least some gay scientists to try to prove that homosexuality is innate. This is an important point because several very controversial studies have been published by openly gay scientists.

Homosexuality and Brain Structures

One of the most controversial studies published in the last several years was a study of brain structure in homosexuals, since this study concluded that characteristic structural differences exist between heterosexual and homosexual men.[89] This study was published in *Science* magazine, one of the most influential and authoritative publications in all of science, by an openly gay neuroscientist. The study was widely publicized by the news media because it made the dramatic claim that sexual orientation could be based, at least in part, on biological differences in a critical brain structure. Yet the study was so deeply flawed that it is arguable whether it should have been published at all, let alone in the hallowed pages of *Science*.

This controversial paper described the results of a series of measurements made of the size of a particular brain structure, called the third nucleus of the anterior hypothalamus. Interest was focused on this brain region for several reasons. Previous work by other scientists had shown that the anterior hypothalamus is involved in the regulation of sexual behavior. Male monkeys with an injury to the anterior hypothalamus have an impaired ability to display heterosexual behavior, yet their sex drive is nonetheless intact, causing them to display sexual behavior to monkeys of the same sex. Other scientists had also shown that, in humans, there is a male–female difference in the size of the nucleus, such that this structure is significantly larger in men than in women.

The difference in size of the third nucleus between men and women could be interpreted in another way entirely: perhaps a small third nucleus is characteristic of individuals sexually interested in men, whereas a large third nucleus is characteristic of individuals who are sexually interested in women. To test this idea, measurements were made in brain tissue from gay and straight men, as well as in brain tissue from straight women, to test the prediction that the size of the third nucleus in gay men would be more similar to that of straight women than to that of straight men. Measurements could not be made in living subjects, since the structures cannot be visualized by magnetic resonance imaging (MRI), and are located in the deep central portion of the brain. Therefore, the measurements had to be made in tissue removed at autopsy from subjects who had died for a range of reasons. Tissue was obtained from 19 homosexual men, 16 men who were presumed to be heterosexual, and 6 women also presumed to be heterosexual. Brain tissue was preserved and sliced into thin sections, to be viewed by microscope. The physical dimensions of all four interstitial nuclei were measured in the anterior hypothalamus of all 41 subjects. It was found that the third interstitial nucleus was the only nucleus that differed in size among the three groups of subjects, and this structure was twice as large in heterosexual men as in homosexual men or heterosexual women. On the face of things, this result tends to confirm the male–female difference that had been previously noted, and it also suggests that the third nucleus of a gay man is structurally more similar to that of a female than that of a heterosexual male brain. However, the range of variation in size of this structure was tremendous, and size classes were broadly overlapping between the groups. Nevertheless, the difference in size was statistically significant.

However, this study was fatally flawed, because of the source of the brain tissue analyzed. Homosexual men were identified as homosexual, not by an in-depth interview, but rather because they were hospitalized for complications of AIDS. In fact, 100% of the 19 homosexual subjects died of AIDS, and these men were identified as homosexual on the basis of a medical history taken at hospital admission. Similarly, the heterosexual men were not classified as such on the basis of an interview; they were merely assumed to be heterosexual because of the numerical preponderance of heterosexuals in the population. Yet, among the group of 16 men assumed to be heterosexual, 38% died of complications of AIDS. This creates several major problems. First, it seems highly unlikely that 38% of a random sample of heterosexual men would die of AIDS; this may mean that some of the "heterosexual" men were in fact homosexual, or it may mean that some of these men were intravenous drug users. In any case, it is not a fair test to compare a group of homosexuals to a group that may be comprised of heterosexuals, closet homosexuals, and intravenous drug-users. Second, 100% of the gay men died of AIDS, whereas only 38% of the "straight" men died of AIDS. The AIDS virus is well known to cause progressive pathology in the brain of many individuals with AIDS. Therefore, it is quite possible that the differences observed in this study were actually related to the effects of the AIDS virus, not to any difference in sexual preference. Patients with AIDS frequently have a reduced level of testosterone in their bloodstream, the result either of the virus or of treatment for the virus, so it is also possible that the reduced size of the nucleus could be caused by a hormone abnormality. When homosexual men in this study were compared to the 6 "heterosexual" men who died of AIDS, the observed size difference in the third nucleus almost disappeared. In summary, although the study may eventually be corroborated, it is simply not convincing in itself. The poorly done comparison between gay and "straight" men should have precluded publication in a preeminent journal like *Science*. The truth of the matter is that, had the study been published in a lesser journal, no one would have paid any attention to it.

More recently, another study identified a difference between gay and straight men in another brain structure. The newer study showed that the size of a part of the brain called the anterior commissure differed depending on sexual orientation.[90] This study also used brain tissue collected at autopsy from people who had died of AIDS and a range of other causes. Again there was a problem, because 80% of the homosexual men had died of AIDS, whereas only 20% of the heterosexual men had died of AIDS, although an effort was made to exclude all persons who showed any evidence of pathology affecting the brain tissue. But AIDS can cause subtle brain pathologies that might not be easily observed, so it is still possible that this is a problem. In any case, it was found that the anterior commissure is 13% larger in women than in heterosexual men. However, in homosexual men, the anterior commissure is 18% larger than in women and 34% larger than in heterosexual men. When a comparison was made between homosexual men without AIDS and heterosexual men without AIDS, it was found that the anterior commissure was still 35% larger in homosexual than in heterosexual men. This discrepancy could not be accounted for on the basis of differences in brain size, since the discrepancy remained when the size of the anterior commissure was corrected for differences in brain size.

The differences between homosexual and heterosexual men in the second study are probably not related to differences in exposure to the AIDS virus. The AIDS virus is more likely to cause atrophy or shrinking of brain structures than it is to cause an increase in size, as was observed here. Furthermore, when homosexuals with AIDS were compared to homosexuals without AIDS, there was no significant difference in the size of the anterior commissure. Similarly, AIDS was not associated with significant atrophy of the anterior commissure in heterosexual subjects. Finally, microscopic examination of several brain tissue samples from AIDS patients showed no evidence of the sort of pathology that would lead to tissue swelling. All of these considerations suggest that the second study is stronger than the first.

These results are especially intriguing because the anterior commissure is not a structure known to be involved in sexual behavior. It is simply a tract of neurons that connects the right and left halves of the brain, which helps the two sides of the brain communicate with one another. There was no reason to suspect that this structure should in any way be related to sexual orientation. In humans, the anterior commissure mediates the transfer of visual, auditory, and olfactory information from one hemisphere to the other. The functional significance of a size difference in this structure is completely unknown, although men and women are known to differ somewhat in how the two sides of their brain function. For example, heterosexual women and homosexual men both tend to score higher on verbal tests and lower on tests of spatial visualization than do heterosexual men. Because it is possible that the size of the anterior commissure somehow determines how the two sides of the brain work together, it may be that there is less "lateralization," or lateral specialization, in the brain of subjects with a large commissure. But, although one can imagine how a large anterior commissure might contribute to better verbal ability in women, it is hard to imagine how a small anterior commissure could be related to better spatial ability in men.

The differences in brain structure between homosexual and heterosexual men suggest that homosexuality is not related to changes in any one brain structure. Rather it may be true that whatever factors originally cause homosexuality also cause a multiplicity of changes in the human brain. It may turn out that there are numerous structural differences between men and women, and between homosexual and heterosexual men. However, the limitation of both this study and the previous study is that structural differences between individuals do not prove that there are genetic differences between individuals. To put it another way, the fact that there are brain differences between homosexuals and heterosexuals does not prove that these brain differences cause the difference in sexual orientation; in fact, the structural differences could conceivably result from whatever made the subjects different in the first place. It seems unlikely that anything as clear-cut and obvious as the volume of a brain structure could be in any way informative about something as ambiguous and subtle as sexual orientation.

Homosexuality and the Modern "Split Twin" Experiment

Years ago it was noted that the brothers of homosexual men are more likely to be homosexual themselves than are the brothers of heterosexual men. However, this sort of observation does not address the issue of whether a familial trait is related to shared environment or to shared genes. This issue can only be addressed by examining men who share the same environment but who differ in the degree of their genetic relatedness. This is, of course, the workhorse of behavioral genetics: the split twin experiment. The strongest approach is to compare identical and fraternal twins with each other and with adopted male siblings, who have a similar environment but little or no genetic relatedness.

Recently a large study was published in which twins were used to determine the importance of shared genes in defining sexual orientation.[85] Male subjects were recruited for the study by placing advertisements in gay publications throughout the country, which probably means that only openly gay men were recruited. The ad specified that subjects should be gay or bisexual men, with either male twins or adopted brothers of similar age. A telephone number was published so that interested subjects could contact the scientists doing the research. All potential subjects were interviewed in depth to verify that the men were appropriate subjects, then scientists made an effort to assess the sexual preference of the subjects' brothers. Sometimes this assessment was made indirectly, by questioning the subject about his brother, but if neither the subject nor the brother objected, the assessment was made in person or by a telephone interview. Interviews were obtained for a total of 115 men with identical or fraternal twins, and another 46 men with adoptive brothers. About 74% of the brothers of these subjects also participated, meaning that information about sexual preference could be obtained directly from the brother. The remaining 26% of brothers did not participate directly, so that information about their sexual preference was obtained indirectly from the subject. It is a major potential weakness that fully one-quarter of the brothers of the subjects did not directly participate in the study.

Nevertheless, it was found that there was a significant and relatively strong effect of genes on sexual preference. More than half (52%) of the identical twins of gay men were also gay, while only 22% of the fraternal twins were gay.[85] Among the adoptive brothers, only 11% were gay, which confirms that there is a strong genetic component to sexual orientation. The fact that the incidence of homosexuality in adoptive brothers is not substantially higher than the population at large suggests that environment is actually a weak determinant of sexual preference. These reported differences were statistically significant, as there was less than one chance in 1000 that the results were arrived at by chance alone. Since this study involved completion of a questionnaire by all subjects, it was possible to compare heterosexual and homosexual men by several different criteria. For example, it was found that male heterosexuals could recall significantly more of their own aggressive behavior as children than could male homosexuals. Heterosexuals also recalled more childhood interest in sports than did homosexuals, and were less likely to be involved in childhood play that was regarded as "sissy" by their friends.

A mathematical model was used to estimate the heritability of homosexuality, or that fraction of homosexuality that is likely to be explained solely on the basis of shared genes.[85] The best estimate is that heritability explains roughly half (54%) of the concordance between twins. Using the same model, it was calculated that shared environment likely explains none of the concordance between twins, and that unshared environment probably explains about 34% of the concordance between twins. Shared environment is that part of the environment that is actually common to both siblings, and probably represents virtually every aspect of home life. Unshared environment, which explained about one-third of the concordance between twins, is perhaps best regarded as that part of the environment external to the home (although, in some cases, the home may include unshared environment even for identical twins). If these results are correct, it implies that the environment is only weakly determining for sexual orientation. Because genetics is so important in determining sexual orientation, the oft-voiced fear that a homosexual teacher may lure his pupils into homosexuality is probably groundless.

Even though this study was good, it has certain weaknesses that could potentially invalidate its conclusions. Only 161 homosexual men were initially contacted for this study, which seems to be a relatively small number of men to examine a topic as complex and subtle as sexual preference. Furthermore, only 74% of the subjects' brothers responded to a questionnaire, so it is possible that the brothers who did participate are not a representative sample of men. There is a very strong possibility that homosexual brothers would be more likely to answer the questions at all, whereas heterosexual brothers might be more likely to answer the questions honestly. Finally, it is possible that a special "twin environment" contributes to the development of same-sex orientation, so that there are inherently more homosexuals among identical twins. This would mean that the estimate of heritability of sexual orientation is too high among identical twins, and that the importance of the environment is underestimated. Yet despite these caveats, the levels of statistical confidence reported in this study are good; there is less than one chance in 1000 that results as strong as these could have been obtained by accident. Or, to put it another way, we can be roughly 99% certain that heredity has a strong impact on sexual preference.[85] These same scientists recently showed that there may also be a genetic basis for homosexuality in female identical and fraternal twins, so that both male and female sexual orientation appear to be strongly influenced by genetic factors.

Some of these results have been independently confirmed by another study that examined sexual behavior among 158 sets of male twins.[87] This study speculated that identical twins may have a more stressful prenatal environment than fraternal twins, as shown by the higher incidence of birth defects and fetal death among identical twins. Since prenatal stress can lead to homosexuality in animals, it is possible that identical twins are inherently more likely to be gay than are fraternal twins. In addition, identical twins are more likely to imitate each other or to cooperate with each other, so it is possible that, in some individuals without a strong sexual orientation, homosexuality may be a behavior learned from the twin. These latter possibilities hint that a "split twin" approach may not be sophisticated enough to investigate something as complex as sexual preference.

Evidence for a Genetic Basis of Homosexuality

Very recently, sophisticated techniques of molecular genetics were used to determine whether there is actually a genetic basis for homosexuality. This is possible because traits leave genetic trails that can be followed by molecular biology, and the pattern of transmittance of a trait from one generation to the next can be studied in great detail. If the pattern of transmittance of a trait across generations is characterized in this way, this can help determine whether or not the pattern is consistent with that of an inherited trait. In many cases, this type of pedigree analysis can yield insight into the genetic mechanisms of inheritance for a particular trait, including the location of a particular gene on a particular chromosome.

If homosexuality is genetically influenced, there should be a significantly higher incidence of homosexuality within certain families than in the general public. To determine if this is correct, a study group of 114 self-acknowledged homosexual men was recruited through two AIDS outpatient clinics and through several gay organizations.[91] A pedigree analysis of these men was performed by asking each man about the sexual orientation of relatives, including father, sons, brothers, uncles, and male cousins. The major weakness of this type of approach is that it is not always possible to determine a person's sexual orientation without asking them, it is not always possible to ask, and some people may not be forthcoming even if they are asked. Thus, some of the relatives in this study might have been incorrectly classified as to their sexual orientation. Each of the men in the study also completed a detailed questionnaire that asked about his gender identification, sexual development, and sexual behavior, as well as a wide range of other questions related to personal and family medical history. The questionnaires revealed that most of the gay men in this study experienced their first attraction to another male by the age of 10, admitted to themselves that they were homosexual by the age of 16, and "came out of the closet" by the age of 22. Since the average age of the men in the study was 36, most of these men had spent essentially their entire adult life as practicing homosexuals. Finally, each of the men in the study gave a blood sample which was analyzed to build a genetic portrait of each study participant.

It was found that homosexuality is indeed strongly clustered in some families; among the brothers of men in this study, the incidence of homosexuality was nearly seven-fold higher than in the population at large. Among more distant relatives, the incidence of homosexuality was about three-fold higher on the maternal than on the paternal side of the family. About 8% of both maternal uncles and maternal male cousins were homosexual, whereas only 2% of the population at large is homosexual according to the criteria used in this study. Because uncles and cousins share genetic information, yet are raised in separate environments, this strongly suggests a genetic explanation for homosexuality. But what is curious is that there were relatively few male homosexuals on the paternal side of the family, and very few female homosexuals on either side of the family. Many scientists might have guessed that homosexuality is genetically determined, but few of these scientists would have expected there to be a pattern suggesting that the trait is passed exclusively through the female line. Such evidence for maternal inheritance of homosexuality is truly startling. Moreover, the idea that male and female homosexuality are somehow inherently different, so that they do not cluster together in the same families, is also quite unexpected.

The fact that male homosexuality seems to be transmitted through the maternal line suggests that inheritance of the trait is somehow linked to inheritance of the X chromosome. Women have two X chromosomes, one contributed from each parent, so that unusual or aberrant genetic traits carried on this chromosome are usually not expressed. This is because even if one of the X chromosomes is aberrant or unusual, the other will likely be normal, and thus the normal version of the trait can still be expressed. But men carry only one copy of the X chromosome, contributed by their mother. Instead of a second X chromosome contributed by his father, each man instead inherits a Y chromosome from his father. This means that men have only one copy of those genes on the X and Y chromosomes; if that one copy of the gene happens to be aberrant or unusual, then there is no option but to express the gene. This means that men are much more likely to express X-linked traits than are women. The only way in which a woman can express an aberrant or unusual X-linked gene is if she inherits the same X-linked mutation from both parents. While this is not impossible, it is far less likely to occur than a man inherit one copy of an X-linked mutation from his mother.

The idea that homosexuality can be passed down from mother to son via the X chromosome is merely a hypothesis; a plausible but unproven explanation of the facts. This hypothesis can be tested by carefully examining the genetic inheritance of both gay and straight men on the molecular level, to see whether the observed pattern of inheritance conforms to predictions of the hypothesis. If the observed and predicted patterns are the same, then the hypothesis may be correct, whereas if the observed and predicted patterns differ, the hypothesis is necessarily incorrect. This type of hypothesis testing can be very difficult to do, since the exact nature of the gene associated with homosexuality is not yet known, and because it is not yet possible to examine DNA directly, to test whether an unknown gene is present. Instead, scientists must test the DNA to determine whether certain gene markers are present. If an unknown gene on the X chromosome causes homosexuality, then all or most gay men should have this gene, as well as having other genes that are physically nearby on the same chromosome. If nearby genes can be identified, they can serve as markers to reveal the presence of the unknown gene for homosexuality. In other words, known marker genes that happen to be linked to an unknown gene can be used to reveal the presence of the unknown gene, in much the same way as fingerprints can reveal the presence of an unknown intruder.

To test the idea that homosexuality is passed down on the X chromosome, gene markers on the X chromosome were examined in a group of 40 homosexual men who also had homosexual brothers. This sample of men was chosen simply because it increased the likelihood of identifying the gene for homosexuality, although it could well be argued that this is a very unusual group of men. Blood samples from these 40 gay men and their brothers were tested, together with blood samples from their mothers and other siblings whenever possible. It was found that one portion of the X chromosome seemed to carry a gene associated with homosexuality; 33 of the 40 gay sets of brothers were concordant for gene markers within this small region. The fact that 7 of the 40 gay brothers were not concordant for gene markers in this region probably means that there are other genetic or nongenetic sources of homosexuality.[91] Of course, these results must be replicated and extended before full faith should be put in them, but this study appears to have been very carefully done.

Genes and Human Sexuality

If homosexuality is indeed genetically determined, it is fair to ask why there should be a gene for a behavior pattern that is, in an evolutionary sense, strongly selected against. On average, gay men have five times fewer children than straight men, even when those gay men who marry and try to live a heterosexual life are included.[85] If straight men, who are numerically dominant in the first place, also have five times as many children per capita as gay men, the proportional genetic contribution of gay men to succeeding generations must diminish over time. Evolutionary fitness, in the sense of Charles Darwin, is proportional to the genetic contribution of an individual to succeeding generations. This means that gay men have low evolutionary fitness, since straight men make a much larger genetic contribution to the next generation. So why does homosexuality persist in the gene pool at all?

Conceivably, "gay genes" could have adaptive value if they predisposed gay persons to assist in the rearing of related offspring. For this to be true, the child care provided by gay men and women would have to strongly increase the chances of survival of their nieces and nephews who share the same genetic heritage. By increasing the survival of these children, homosexuals would indirectly increase their own proportional contribution to the next generation. Yet this seems to be patent nonsense; for this type of indirect effect to be true, the nieces and nephews of gay men would have to enjoy a much higher rate of survival than normal, and this is not true. This type of indirect benefit of "gay genes" would also imply that child care should be a major component of gay culture, which it apparently isn't. Although some gay men and women rear their own children or even adopt other children, this is the exception rather than the rule. Nevertheless, the fact that homosexuality persists as a trait, despite putting homosexuals at an apparent evolutionary disadvantage, suggests that homosexuality is, in some sense, an adaptive trait.

Perhaps the putative "gay gene" confers some subtle survival advantage to women who bear the trait but are not themselves homosexual. If women who carry the homosexual gene are somehow able to bear more children over time, this type of evolutionary advantage could offset the evolutionary disadvantage experienced by men who express the homosexual gene.[92] This hypothesis of "balancing selection" could be tested by determining whether the sisters of gay men have more children on average than the sisters of straight men. Alternatively, the gay gene may be created by mutation at a rate high enough to offset the loss of old mutations from the gene pool. Perhaps the X chromosome is somehow unstable and particularly likely to mutate, so that new mutations of the gay gene are commonplace. This hypothesis could be easily tested, because it would imply that the structure of the X chromosome should be somehow different in diverse gay men. However, it should be clearly stated that it is sheerest speculation that the gay gene, if it even exists, confers some sort of selective advantage.

Several recent studies do strongly suggest that there is at least one gene whose gene product is associated with homosexual behavior. It may be that the gene product made by the X chromosome is a hormone or hormone receptor, which could have an effect during a certain developmental window, or it may be that the nature of this gene product is completely unknown. Although no mechanism is known for how a gene product could affect sexual behavior, this deficit in our understanding does not argue against the finding itself. In many cases, an observation was made long before scientists were prepared to offer an explanation for the observation. As an example, bumblebees can fly, even though theorists can prove conclusively that they shouldn't be able to get off the ground.

12

Alcoholism and Addictive Behavior

Alcohol use in the United States is exceedingly common, yet half of all the alcohol drunk is consumed by only 10% of the population.[93] This in itself implies that some individuals drink too much. Yet consuming "too much" alcohol does not make one an alcoholic, and there is great variation from place to place, and from one era to another, in what is considered acceptable. To further complicate the picture, alcohol consumption varies between the sexes, between socioeconomic strata, and between occupations. Consequently, it is very difficult to determine what is "normal" for alcohol consumption. Yet it is often rather easy to identify someone who exceeds that normal level of consumption.

The American Psychiatric Association defines alcoholism as being of two types. The less damaging type is alcohol abuse, which is a psychological dependence on alcohol. An alcohol abuser is psychologically dependent on alcohol, he may indulge in occasional heavy alcohol consumption, and he continues to drink despite mounting evidence of occupational or social problems. The more damaging type of alcoholism is alcohol dependence, which is both physical and psychological dependence on alcohol. An alcohol-dependent person shows all of the signs of alcohol abuse, together with signs of increased alcohol tolerance and physical symptoms on withdrawal from alcohol. It has been estimated that between 10 and 31% of men in the United States are alcoholic, so prevalence of the disease is quite high.[10]

Virtually everyone has firsthand knowledge of alcoholism, either as the relative or close friend of an alcoholic, or as one who suffers from the disease. It will come as no surprise to most that alcoholism is familial, and that a tendency to alcoholism severely affects some families. But whether such a family history is caused by shared genes or shared environment has been a contentious issue for a long time.

The idea that drug addiction is similar to alcoholism, and that addictive behavior in general may be hereditary, is more controversial than the idea that alcoholism is hereditary, in part because we know so much less about drug addiction. But alcoholism and drug addiction have much in common, and both types of addictive behavior can be devastating. Alcohol is believed to have been the underlying cause of death for as many as 100,000 people in the United States in 1990 alone.[94] This number is so large because alcohol is thought to contribute to 40–50% of all motor vehicle fatalities, 16–67% of all accidental deaths in the home and workplace, 60–90% of all deaths from cirrhosis of the liver, and 3–5% of all cancer deaths. By contrast, illicit drugs of all kinds, including crack and heroin, even when combined with drug-related accidental deaths, were responsible for about 20,000 deaths in the same year. Thus, from a national perspective, alcoholism is a far more serious problem than is drug abuse, since alcohol causes five times as many deaths.

Split Twin Studies of Alcoholism

The children of alcoholic parents are about five times more likely than normal to become alcoholic themselves.[10] But this does not prove that alcoholism is hereditary, because children of alcoholic parents may have been brought up in an environment that encourages alcoholism. As we have seen so often before, the most reliable first approximation of the heritability of alcoholism is obtained by studying twins split at birth and raised in different environments. This was done recently, using the records of several large alcohol abuse treatment centers in Minnesota. A search through the medical records identified 599 patients who had received treatment for alcohol abuse and who reported having a sibling of the same age. Both siblings were contacted, to determine whether they were twins and to recruit them to the study if possible. Twins were eliminated from the study if both could not be contacted, or if both were not of the same sex, since males and females may not have the same risk of alcoholism. An exhaustive effort enrolled a total of 169 same-sex twin pairs, of whom roughly half were identical, half fraternal. There were more than twice as many male twins than female twins, which is consistent with the higher incidence of alcoholism among men. But there were still a fair number of female twins enrolled, making it possible to determine whether men and women indeed do differ in their risk of alcoholism.

It was found that identical twins were at a higher risk of concordance for alcoholism than were fraternal twins, which clearly shows a role for genes.[10] However, the difference between identical and fraternal twins was fairly small, suggesting a rather weak role for genes in the genesis of alcoholism. For males, identical twin concordance was 76%, while fraternal twin concordance was 61%. A computer analysis of these concordances concluded that alcoholism is only 36% heritable for men. For females, identical twin concordance was 36%, while fraternal twin concordance was 25%, implying that heritability for women is only 26%. These conclusions are consistent with the impression that alcoholism risk is greater for men than women, but the overall risk of alcoholism related to genes alone is surprisingly small.

Interestingly, it was found that, for both men and women, alcohol dependence is substantially more heritable than alcohol abuse (Table 1). From a psychiatric standpoint, alcohol dependence and alcohol abuse are easily separated from one another. Dependence tends to develop quickly in certain teenagers and young adults, whereas abuse develops .slowly in middle-aged and older people. Dependent alcoholics often describe becoming dependent very soon after their first drink, whereas alcohol abusers typically were able to drink socially for many years before problems developed. Alcohol-dependent people show alcohol-seeking behavior and often have an antisocial personality type, whereas alcohol abusers tend to be anxious, especially

Table 1
Characteristics of the Two Types of Alcoholism[a]

Characteristic features	Type of alcoholism	
	Abuse	Dependence
Age at onset	>25 years	< 25 years
Ability to abstain for extended periods	Common	Rare
Social drinking to overcome anxiety	Common	Rare
Guilt and fear about alcohol dependence	Common	Rare
Fighting and arrests while drunk	Rare	Common
Alcohol-related job problems	Rare	Common
Multiple drinking binges	Rare	Common
Alcohol needed to maintain function	Rare	Common
Probable cause of alcoholism	Loss of control	Alcohol-seeking
Basic personality type	Anxious	Antisocial
Heritability		
For men	38%	60%
For women	0%	42%

[a]Adapted from several studies comparing alcoholism in twins.[10,93]

in social situations, and to drink excessively because they lose control. To put it another way, alcohol dependence appears to be a manifestation of an underlying personality disorder, whereas alcohol abuse may be a way to cope with normal levels of stress. In this light, it is not surprising that familial risk is highest for early onset alcoholism, a type of dependence that almost certainly arises from a personality disorder and can develop in very young teenagers. This type of alcohol-seeking behavior is an entity quite distinct from alcohol abuse, although abusers may inherit a susceptibility to loss of control after drinking. The familial risk of alcoholism is higher for the children of severe or early onset alcoholics, so a severely affected parent is more likely to have alcoholic children than is a mildly affected parent. The fact that level of risk varies in this manner implies that alcoholism is a multifactorial disorder, caused by several to many different genes acting somehow in concert.

The major weakness of this study is that it examined a group of people with an alcohol problem severe enough to drive them to seek treatment. Most alcoholics never seek treatment, so the twins in this study were probably more severely affected than most alcoholics. Generally speaking, those alcoholics who do seek treatment are more likely to have a range of psychiatric problems associated with their illness. The twins analyzed in this study had a higher-than-average incidence of clinical depression and anti-social personality disorder, and also tended to have an early onset of alcoholism. In addition, these twins were almost exclusively white, so that little insight could be gained into the heritability of alcoholism in other races. Finally, there may have been a problem in the mathematical model used to calculate heritability. This model assumed that the parents of an alcoholic child are each no more likely than average to be alcoholic themselves. Of course, this assumption would fail if an alcoholic chooses to have children with another alcoholic. In other words, if the mating options of an alcoholic are at all constrained by their disease, as they would likely be, then alcoholics may be more likely than normal to have children with another alcoholic. This kind of "assortative mating" can cause major problems for any mathematical modeling study of inheritance. For these several reasons, it is unclear to what extent the findings of this study can be generalized to the population at large.

In any case, the role of the environment appears to be considerably stronger than the role of the genes in addictive behavior, for men and women both. In fact, environment is almost half again as important as genes, for everything except alcohol dependence. If alcohol dependence is regarded as a kind of chemical imbalance that drives the individual to seek alcohol, it is not surprising that this is strongly genetic. Similarly, if alcohol abuse is regarded as social drinking gone malignant, it is also not surprising that this is strongly environmental. The varying importance of genes in the causation of alcoholism is

shown clearly by the fact that the heritability of alcoholism
varies as a function of the age of onset of illness in male twins.[95]
It was found that if the onset of problem drinking occurred at or
before age 20, the heritability of alcoholism was as high as 73%,
whereas problem drinking with a late age of onset was only 30%
heritable. Men with an early onset of problem drinking were also
more at risk for drug use, school misconduct, and precocious
sexual behavior.

The environment clearly has a strong causative role in
alcoholism, as confirmed by cross-fostering studies.[93] A cross-
fostering study is one that follows children from an alcoholic
familial background, who are adopted by either alcoholic or
nonalcoholic parents. Any differences in the incidence of alcoholism in these children can be attributed to the environment, since
all of the adopted children share a similar genetic heritage.
Similarly, children from a nonalcoholic background, adopted by
either alcoholic or nonalcoholic parents, are followed, to round
out the picture. This kind of cross-fostering study is quite
difficult to do, since adoption agencies typically try to place
adopted children with families that are not affected by alcoholism. Yet even the best agencies make mistakes, and alcoholism
may develop as a problem after a child has already been placed
with the adoptive family. Cross-fostering studies are generally
considered to be a very robust way to address heritability, since
the influence of genes can be clearly separated from the influence of the environment.

Scientists working in Sweden were able to find 862 adult
men, all born to single women in Stockholm between 1930 and
1949, and all of whom had been adopted at a young age by
nonrelatives.[93] Complete information about alcohol abuse, mental illness, and medical illness was available for each of these
adoptees, and for their biological and adoptive parents, because
of the sophisticated public health system in Sweden. Both
children and parents were classified as to whether they were
alcohol-dependent or alcohol abusers, on the basis of these
medical records. Alcoholics were considered to be abusers if they
became alcoholic after the age of 25 and if there was no record of

Table 2
Rate of Alcoholism among Adopted Children[a]

| | Abuse | | Dependence | |
| | Environmental exposure | | Environmental exposure | |
Genetic exposure	No	Yes	No	Yes
None in family	4%	4%	2%	4%
Family history	7%	12%	17%	18%

[a]The rate of alcoholism in adopted male children, coming from different genetic backgrounds and adopted into different environments. These data are from a cross-fostering study design, in which the inheritance of severe alcoholism is determined in children adopted away from their biological parents.[93] This study design enables scientists to separate hereditary and environmental causes of alcohol abuse or dependence.

imprisonment or criminal behavior. Alcoholics were considered to be dependent if their alcoholism began in adolescence, and if there was a record of serious criminality beginning at about the same time. Results from this study showed clearly that there is a strong component of both environmental and genetic causation in alcoholism (Table 2).

Children with no family history of alcoholism are generally at a low risk of becoming alcoholic themselves, no matter what the environment is like. Children with a family history of alcohol abuse are at a somewhat higher-than-normal risk of becoming alcohol abusers themselves, even if the adoptive family environment does not include an alcoholic. But these children are at a much higher risk of abuse in an alcoholic adoptive family. Children with a family history of alcohol dependence are at high risk of alcohol dependence, whether or not the family environment includes an alcoholic. Overall, it was found that a family history of alcohol abuse increases a child's risk of alcoholism about twofold, while a family history of alcohol dependence increases risk fivefold. On the other hand, an alcoholic family environment does not increase the risk of alcohol abuse at all in the absence of a family history. An alcoholic family environment does increase the risk of alcohol dependence about twofold, even in the absence of a family history of alcoholism, but this finding is based on a very small number of children at risk. Overall, the effect of low social status is to increase the risk of alcohol abuse for all people. The greatest weakness of this study arises from the great strength of the adoption system in Sweden; nearly four times as many children were placed with nonalcoholic adoptive parents as were placed with alcoholic parents. Thus, relatively few children were put at risk because of an alcoholic family environment. Consequently, scientists may not have an accurate assessment of the role of environment in the causation of alcoholism.

The Sobering Story of the A1 Allele

Because alcoholism is a disease that has many manifestations and is clinically complicated, one might expect it to be fairly complicated from a genetic standpoint. Yet this expectation was completely overturned by a group of scientists who found evidence that one particular gene, called the A1 allele (or form) of the dopamine receptor, was strongly associated with alcoholism.[96] Dopamine is a chemical that acts as a messenger between neurons in the brain; the dopamine receptor is the protein that actually binds to dopamine and so receives the signal carried by this molecule. It was intuitively appealing to many scientists that the dopamine receptor be involved in addictive behavior because dopamine itself is known to be involved in some pleasure-seeking and addictive behaviors. Consequently, a gene that controls expression of the dopamine receptor would be an obvious site of mutation in alcoholics. Yet this study remains largely unreplicated to this day, even though a great deal of effort has been spent trying to confirm the findings. Several of the original scientists have backed away from the claim that the A1 allele is critical in alcoholics, and most scientists now believe that the original report was wrong. But the story of the A1 allele remains as a sobering demonstration of the difficulty involved in doing behavioral genetics.

The original data about the A1 allele of the dopamine receptor was obtained by scientists who studied preserved brain tissue from a tissue bank.[96] Tissues were obtained from 35 alcoholics and 35 nonalcoholics, all of whom had died many years previously, and who were then paired by age, sex, and race. Each of the tissues was tested for differences in expression of nine different genes. The only gene found to differ significantly between the two groups was the dopamine receptor allele known as A1. The A1 allele was present in 69% of the alcoholics, but in only 20% of the nonalcoholics. Given that the number of samples analyzed was reasonably large (70 in all), this difference was highly significant from a statistical standpoint. Although no one knew how a single protein could predispose an individual to develop a disease as complex and variable as alcoholism, the finding nevertheless implied that the dopamine receptor was somehow linked to development of alcoholism. This was very exciting, because it raised the possibility that a test could be developed to screen for alcoholism in the general population. Since 28 million Americans are the children of alcoholics, a strong linkage between alcoholism and the expression of a single gene would make it quite easy (and very lucrative) to develop a test kit to identify those most at risk of alcoholism. Initially, the only reservation about this study was that 20% of nonalcoholics also had the A1 allele; this makes a test kit less useful, since many people identified as being at risk would never actually become alcoholic. But this finding also seemed to imply that the A1 allele is not strongly linked to alcoholism, since a fair number of people with the gene never did develop the illness.

Soon after publication of the first piece of research on the A1 allele, another study was published that confirmed that the A1 allele is more common among alcoholics.[36] However, the newer study also suggested that the A1 allele might be involved in a range of mental illnesses, including attention deficit–hyperactivity disorder, autism, and Tourette's syndrome (Tourette's is an unusual mental illness that causes the sufferer to have uncontrollable outbursts of repetitive behavior, including verbal and physical tics, incomprehensible speech, and violent swearing). This newer study of the A1 allele analyzed brain tissue removed at autopsy from 108 healthy individuals, and compared this to tissue from 86 alcoholics, 48 patients with attention deficit–hyperactivity disorder, 18 autistic patients, and 147 patients with Tourette's disease. The A1 allele was present in 22% of the normal individuals, but it was found in 43% of alcoholics, 46% of patients with attention deficit–hyperactivity disorder, 55% of autistic patients, and 45% of Tourette's patients. Therefore, incidence of the A1 allele was more than twice as high as normal in all four of the patient groups. This finding was significant in a statistical sense, but relatively unimpressive nonetheless, because it shows that the A1 allele is not closely linked to alcoholism. Since the A1 allele was present to the same extent in several different disease states, some of which are not at all closely related to alcoholism, the linkage specifically to alcoholism is weakened. Yet the most striking thing about the newer study was that no patient group had the very high frequency (69%) of the A1 allele found in the first study. The fact that the A1 allele is not found in most alcoholics implies that it is not the primary cause of alcoholism, although it may act to modify the expression of another gene more closely linked to alcoholism. Perhaps the A1 allele is more common among alcoholics who develop medical problems from their drinking, or perhaps the A1 allele is somehow associated with a range of different compulsive diseases including alcoholism.

More recently, the A1 allele has fallen even further from grace. A study that examined 44 alcoholics and 68 nonalcoholics was unable to find any difference in A1 allele frequency between the two groups.[97] When alcoholic subjects were more closely examined, it turned out that A1 allele frequency was unrelated to age of onset, family history, total alcohol consumption, antisocial personality, or physical ailments associated with drinking. A meta-analysis, which is a formal reanalysis of data published by other scientists, showed that there was much more difference in A1 allele frequency between different ethnic groups than there was between alcoholic and nonalcoholic subjects from a single ethnic group.[98] For example, A1 allele frequency is less than 10%

in Yemenite Jews, is about 40% in American blacks, and is about 80% in Cheyenne Indians. Yet the mean A1 allele frequency in alcoholics of all ethnicities is 42%, while the mean allele frequency in nonalcoholics is 34%. When the early data, the data that started the whole controversy, were excluded from further consideration, the mean A1 allele frequency was 36% in both alcoholics and nonalcoholics. Finally, when modern molecular biological techniques were used to examine the fine structure of the A1 allele, it was found that there was no difference between alcoholics and nonalcoholics.[99] A group of 113 alcoholics was compared to a group of 34 nonalcoholics and a group of 106 patients with schizophrenia, and it was found that there were no consistent differences in the A1 allele between any of these groups. In fact, two of the scientists who had reported the first very controversial results were also involved in this last study, thus recanting their belief in the importance of the A1 allele in alcoholism.

The sorry saga of the A1 allele is a vivid demonstration of the fact that gene linkage studies are very difficult to do properly. There is no reason to think that the original paper on the A1 allele was in any way dishonest, but it does appear to be completely wrong nonetheless. Simply because a result is consistent with expectations, and can be explained from a theoretical standpoint, does not make it correct. A new result, even if it seems to have great explanatory power, should not be accepted at face value until it has been replicated. It is incumbent on scientists to test and replicate all new findings, so that grains of truth can be found and the chaff of error discarded. Yet we must be careful not to discard or downplay what we do know about alcoholism; alcoholism is an illness for which many people inherit a genetic proclivity. While we do not know exactly which genes are responsible for this inborn proclivity, we do know that the genes exist.

Is Drug Addiction a Heritable Trait?

Drug addiction is in many ways similar to alcoholism, so there is an expectation that drug addiction will also have a component of heritability. However, there is much less evidence for the heritability of drug addiction than for the heritability of alcoholism, simply because drug addiction has been a serious problem for much less time. A diagnosis of childhood hyperactivity disorder predisposes an adult to become an alcoholic or a drug abuser, suggesting that there may be a set of genes that generally predisposes an individual to develop addiction later in life.[95] Alcoholism and drug addiction also have in common that both are more common among individuals with antisocial personality disorder, i.e., the personality disorder that is associated with pathological disregard for the rights and feelings of others. Both alcoholism and drug addiction tend to afflict those with an inability to inhibit impulsive behavior and little inclination to consider personal risk when weighing behavioral options. There is evidence from experiments with animals that both alcoholism and drug addiction are more easily induced in animal strains that have an abnormally low level of fear or a great interest in novel stimuli[100]; this suggests that alcohol dependence in humans might be linked to similar personality traits. There is even some thought that mapping the genes involved in addictive behavior in mice might eventually allow an extrapolation to humans, but this prospect is, at best, a long way in the future.

The expectation that drug addiction is a heritable trait is at least partially confirmed by the fact that alcoholics are often drug abusers as well. Among male twins separated at birth, at least one of whom was alcoholic, more than 40% of men met the criteria for abuse or dependence on drugs (other than alcohol or tobacco).[10] These findings are consistent with those of other scientists, who found an increased rate of drug abuse among the children of alcoholics, and also among the children of parents with other psychiatric problems. For males, identical twin concordance for drug abuse was 63% and fraternal twin concordance was 44%, while for females, identical twin concordance was 22% and fraternal twin concordance was 15%. It was estimated that drug abuse or drug dependence was 31% heritable for men, but only 22% heritable for women. The higher heritability of drug addiction in men than women is consistent with the higher heritability of alcoholism in men, although the reasons for this sex difference remain unknown. Yet overall, drug abuse is substantially less heritable than alcoholism. In fact, the heritability of drug addiction in women is so low that it is not significant in a statistical sense, so it may not even be real. It is too bad that this study was weakened by the fact that it specifically excluded addiction to tobacco, one of the most powerful and harmful drugs ever used.

The same scientists who first proposed that the A1 allele of the dopamine receptor gene is related to alcoholism recently proposed that this same gene is also related to drug addiction.[101] These scientists reported that the A1 allele was present in 51% of cocaine users but in only 16% of drug-free control subjects. Overall, the gene was found in 44% of 504 drug abusers, while it was present in only 28% of the control individuals. Yet it is not at all clear why the frequency of the A1 allele should vary from 16 to 28% in control individuals who are presumably comparable. While these scientists claim that the new findings are evidence for the A1 allele being somehow involved in reinforcing compulsive behavior, the jury is still out on this claim. Given the recent and repeated failures to replicate the A1 allele data for alcoholism, one cannot view these recent claims with much optimism.

Genes and Addictive Behavior

The relationship between genes and addictive behavior is particularly contentious because several recent high-profile studies were found to be flawed. These studies claimed a strong linkage between alcoholism and a gene known as the A1 allele of the dopamine receptor. The fact that this linkage has now been discredited should not be seen as a failure for behavioral genetics, but rather as a measure of the overall success of science. The A1 allele was apparently a false trail that many scientists were led down, but science is inherently a self-correcting process that involves building consensus. Later, when other scientists were unable to replicate the first study of the A1 allele, these scientists were still able to publish their contradictory findings. The result was a spirited dialogue among scientists which has not yet been completed. Basically, the putative identification of the A1 allele was never seen as the end of the story; it was just the beginning of another chapter in the story.

From split twin studies, there is legitimate reason to think that genes play a role in addictive behavior, whether or not we have yet identified the particular genes involved. The split twin study summarized here involved 169 same-sex twin pairs; certainly it would have been more compelling if more twin pairs had been involved in this study, but 169 is a very respectable sample size. In short, there can be no question that alcoholism, and probably drug addiction as well, has a hereditary component. The only argument is about the relative magnitude of the hereditary component, compared to the undeniable environmental component in addictive behavior.

13

Crime and Violence

Crime and violence are facts of life for many Americans; in 1990 alone, 19 million crimes were committed, and nearly a third of them involved violence.[102] The most common form of violent crime is assault, the attack of one person on another. Aggravated assault, which involves a weapon or causes serious but nonfatal injury, accounted for roughly 30% of all violent crimes, while simple assault accounted for another 50% of violent crimes. By comparison, forcible rape and murder are rare; rape accounts for about 2% of violent crimes, while violent crime results in the death of the victim in 0.4% of cases. Yet, in 1990, 23,000 people were murdered; the fact that these deaths represent less than 1% of the violent crimes in the United States shows the magnitude of the problem. The rate of violent crime in the United States far exceeds that of any other industrialized nation. Among 16 such nations surveyed in 1988, the United States had the highest rate of murder, the highest rate of assault, and the highest rate of sexual assault.

Those most at risk of suffering from violent crime are racial or ethnic minorities; blacks were 41% and Hispanics 32% more likely than whites to be victimized.[102] The rate of death by homicide among blacks is about 5 times higher than among whites, and the death rate among young black males specifically is 20 times higher than among older white females. In many ways, the victimizers resemble the victims; perpetrators of violent crime are overwhelmingly male (89% of those arrested) and blacks are 6 times more likely than whites to be arrested for violent crime. In general, victimizers know their victims before a crime is committed. Perpetrators were acquainted with their victims in the majority of simple assaults, forcible rapes, and homicides, in about 38% of aggravated assaults, and in about 26% of robberies.

Violent crime is a fairly egalitarian crime, in that it is occasionally committed even by those with no prior criminal record, and most career criminals with a long history of nonviolent crime tend to have at least one violent crime on their record. In fact, homicide detectives have learned that anyone is capable of murder, given the right circumstances. Yet a study that tracked a cohort of 10,000 males, born in Philadelphia in 1945, found that during a 27-year follow-up period, 6% of the cohort was responsible for committing 71% of the homicides.[103] The same tiny fraction of men was responsible for about 73% of the rapes and 69% of the aggravated assaults. This shows very clearly that the vast majority of violent crimes are committed by a tiny fraction of the populace.

The fact that most violent crime is committed by a small fraction of people implies that the problem of violence should be easy to solve. Yet this has not been so. While the average prison term served per violent crime roughly tripled in length between 1975 and 1989, the violent crime rate was stable during this period.[102] The probable length of incarceration therefore has relatively little deterrent effect on the violent criminal. In fact, it has been calculated that a 50% increase in the *probability* of incarceration would likely prevent twice as much crime as a 50% increase in the *length* of incarceration. In simple terms, this means that if more criminals could be caught and jailed the crime rate would decrease, even if the length of incarceration did not increase. Because deterrence seems to be relatively ineffective in reducing crime, many criminologists have concluded that the most effective strategy to reduce violence in the long term is to prevent it, rather than simply to respond harshly to it.

Reducing violent crime is a compelling societal goal because so many lives could potentially be saved. Moreover, a mere 1% reduction in violence would likely save society about $1.2 billion, in terms of medical and court costs, lost wages and productivity, and personal damages.[103] Yet it has been difficult to develop strategies for the prevention of violent crime, since this issue is often seen as inherently racist. For example, in 1992, the National Institutes of Health (NIH) proposed sponsoring a conference on "Genetic Factors in Crime," but the conference was canceled because of a strong outcry from black politicians and academics. Their contention was that such a conference would legitimize the view that blacks are genetically more prone to violent crime. Similarly, an initiative by the NIH to fund more research into the causes of violence met with opposition from exactly those ethnic groups that stood the most to gain from a reduction in violence. It has become politically correct to pretend that our society is not deeply troubled by violence, because research into the origins of violence can be seen as a way to lay the blame on a particular ethnic group.

Crime and Violence Are Strongly Related to the Environment

The fact that the United States is so much more violent than Canada, to which it is demographically similar, implies that violence cannot be explained solely on the basis of genes. Furthermore, there has been a dramatic increase in the rate of violent crime in the United States in the past few decades. This period of time is far too short for there to have been a corresponding increase in the frequency of genes that might predispose to violence. It is quite simply impossible that the rapid increase in violent crime is the result of an equally rapid increase in the frequency of genes related to violence, since significant changes in gene frequency usually take dozens, or even hundreds, of generations.

There can be no question that the human tendency to violence is associated with, even inflamed by, an inhospitable environment. Violent crime is associated with poverty, probably for the same reason that mental illness and child abuse are associated with poverty. Poverty is highly stressful, especially when combined with racism and a manifest inequity in the distribution of wealth within our society. The rate of child abuse is 6 times higher than normal in families with an annual income below $15,000.[102] A child born in poverty is much more likely to be abused, much more likely to experience a household with marital violence, and much more likely to grow into a violent adult. It is not at all surprising that poverty, child abuse, marital violence, and violent crime all afflict the same segments of society. And all of these social ills may be caused by the fact that poverty increases the stress of day-to-day life so very dramatically.

The cycle of violence is generally well understood; violence begets violence, and one of the best predictors of future violence is exposure to past violence.[104] Abuse or neglect of a child dramatically increases the risk that the child will become delinquent as an adolescent, and violent as an adult. Yet only 26% of abused children have a record of juvenile offenses and the vast majority of abused children do not grow up to be violent criminals; clearly, other factors are involved. It is known that sex, race, and age are all more accurate indicators of who will commit a violent crime than is childhood abuse. In fact, the best single predictor of violence is maleness. The relationships between sex, race, and age are interwoven in a complex way; boys who become violent criminals are more than five times as likely as girls to have suffered abuse and neglect, and blacks who become violent criminals are more than three times as likely as whites to have suffered abuse and neglect. A black male child who suffers abuse and neglect is far more likely to grow into a violent criminal than is a white female child who does not so suffer, but we cannot know whether this criminality is related to sex, race, or abuse. Thus, the fact that a black male child is more likely to become a violent criminal says nothing at all about a genetic tendency to violence.

Violence is a crime of opportunity, as well as of inclination. This is shown clearly by two facts: violence is usually perpetrated on a close family member; and the availability of handguns increases the likelihood that violence will be lethal.[105]

These important conclusions come from a study done to determine which factors best predict who will become a murder victim. The study analyzed all of the murders committed in three different counties in the United States, each of which includes a large city (Seattle, WA; Memphis, TN; and Cleveland, OH). A total of 1860 homicides were examined, with analysis involving a review of police records, an interview with someone close to the murder victim, and another interview with someone from the same neighborhood, of similar age, race, and sex, who served as a basis for comparison. Nearly 83% of all murder cases analyzed were solved, so the relationship between the murderer and the victim could be established. About 51% of all homicides occurred in the context of a quarrel, argument, or romantic triangle, while another 22% of murders were felony-related, and only 8% were drug-related. In 41% of cases the killer was a spouse, lover, or close relative of the murder victim, while in another 31% of cases the killer was a friend or acquaintance of the victim. Only 4% of murders were known to be committed by a stranger; murder victims were nearly nine times more likely to be killed by a friend or acquaintance, more than eight times more likely to be killed by a spouse or lover, and almost three times more likely to be killed by a close relative. Even if all unsolved murders are attributed to strangers, a victim is more than three times as likely to have been murdered by someone intimate with them than by all other killers combined. This suggests that murder is often committed impulsively, by someone close to the victim who probably never planned to kill.

Most murders are committed using whatever lethal weapon comes easily to hand. The most common murder weapon identified in this study was a gun, as 43% of all murders involved a handgun (strangely, only 6% of murders involved a rifle or shotgun).[105] The next most popular lethal weapon, a knife or sharp instrument, was responsible for 26% of deaths. That handguns were involved in so many more deaths than knives is probably because knife attacks are less often lethal; firearm violence is at least 12-fold more likely to result in death than an assault with another weapon. Overall, a gun in the home increased the risk of murder for house occupants nearly 3-fold over homes that did not have guns. Lethal violence also tends to occur in homes that have a history of alcoholism and violent behavior; if alcoholism caused a work-related problem for any member of the household, the risk of lethal violence increased 11-fold. If any family member received medical attention for past violence in the home, the risk of homicide of a family member increased more than 10-fold. In short, most homicides are impulsively committed, although they may be part of an ongoing pattern of family violence. Death often results because a handgun has created an opportunity for violence to become lethal.

That homicide is a crime of opportunity, as well as a crime of violence, is confirmed by the fact that handgun control reduces the homicide rate.[106] A study compared the rate of violent crime in Seattle, Washington, and Vancouver, Canada, between the years of 1980 and 1986. These two cities were chosen because they are very similar in many ways; both cities have a similar history, similar geography, similar climate, similar access to popular culture through television, similar racial makeup, and similar standards of living. However, Vancouver has a gun-control law while Seattle does not, so handgun access is easy in Seattle. The two cities have very similar rates of burglary, robbery, and simple assault, showing that there is an equivalent criminal component in both cities. But, because of the availability of guns, the number of aggravated assaults was sevenfold higher in Seattle than in Vancouver. Not surprisingly, the risk of death from homicide was almost twofold higher in Seattle. Virtually all the excess risk of death could be explained by a fivefold higher risk of being murdered with a handgun in Seattle, since the risk of homicide by other means was comparable in the two cities. An obvious conclusion of this study is that restricting handgun access would likely reduce the rate of homicide in the United States. However, another legitimate conclusion is that violence is actually fostered by the environment (in the form of handgun availability).

Suicide is also a violent crime that is strongly affected by one's environment.[107] A major study examined all suicides that occurred in Memphis and Seattle over a 32-month period. For each suicide victim, a close relative was interviewed as a proxy, to speak for the victim who could no longer speak for himself, and to characterize the mental and physical state of the victim prior to the suicide. The victim was then compared to someone from the same neighborhood of similar age, race, and sex. During the study period, more than 800 suicides occurred, so there was plenty of data to analyze. It was found that 58% of all suicides involved a firearm. Suicide victims were more likely to have lived alone, more likely to have been taking medication for mental illness, more likely to have been arrested, more likely to have abused drugs or alcohol, and more likely to have been poorly educated. Overall, the best predictor of suicide risk was a previously diagnosed mental illness, since people taking medication for mental illness were 36-fold more likely to commit suicide. But, when all of these variables were taken into consideration, the presence of a gun in the home still increased the risk of suicide nearly fivefold. Given the increase in suicide risk associated with availability of a handgun, and the scant likelihood that such a gun will actually be used for self-protection, it was calculated that a handgun is 37-fold more likely to be used for suicide than for self-protection. This shows clearly that easy availability of a handgun results in a substantial increase in the risk of suicide. Suicide, like homicide, is often an impulsive act, far more likely to occur in an environment that makes it possible to act, quickly and decisively, on impulse.

Criminality Is Heritable

Childhood antisocial behavior is heritable, and strongly predisposes an individual to juvenile delinquency and to adult criminality. But this is not an accurate indicator of who will become a criminal, since 87% of American adolescents participate in some form of antisocial behavior (*e.g.*, assault, vandalism, theft, arson) which they eventually cease before becoming 21.[103] Some evidence suggests that those adolescents who start earliest are most at risk of continuing; boys who were arrested by age 14 were nearly 18-fold more likely to become chronic offenders than those who were not arrested, and chronic offenders are more than 14-fold more likely to commit violent crimes. But adolescent participation in antisocial behavior is a poor predictor of later violent behavior, since at least 65% of those who are arrested while young never go on to commit violent crimes.

Evidence from a large study of adopted children shows that there is a tendency for children to reenact the criminal behavior of their biological parents.[108] A study, which examined a group of 14,427 adoptees in Denmark, compared the arrest record of adoptees to that of their biological and adoptive parents. About 14% of the adopted sons were convicted of some crime even if neither the biological nor the adoptive parents had an arrest record; this percentage is somewhat higher than expected in a random sample of the population, but not strikingly so. If the adoptive parents had been convicted of a crime, but not the biological parents, about 15% of the adopted sons were convicted. Thus, the conviction record of the "environmental" parents has a rather small effect on adopted children. However, if the biological parents had been convicted, but not the adoptive parents, 20% of the adopted sons were also convicted; this shows that the genetic parents have a much stronger impact on the child than do the environmental parents. If both biological and adoptive parents had been convicted of a crime, then 25% of the adopted sons were also convicted. But there was a further subtlety to the data; it was found that there was no heritable tendency to violent crime, although there was a heritable tendency to property crime. Even if a biological parent was convicted of three or more violent crimes, the adoptee was no more likely than normal to be involved in a violent crime. But if a biological parent was involved in three or more property offenses, then the adoptee was twice as likely as normal to commit a property crime. About 4% of the male adoptees accounted for more than 69% of all of the convictions, showing that most crimes are committed by repeat offenders. Perhaps most intriguing of all was the finding that an adoptee was more likely to commit a crime if the biological mother was a criminal, than if the biological father was a criminal. There was a general tendency for crime to be more common among men, suggesting that if a woman becomes a criminal, she may have a very strong genetic compulsion to crime. In the aggregate, these findings suggest that it should be easy to predict who, among a cohort of young adults, is most likely to commit a crime. However, this did not prove to be true. If one considers only those adoptees whose biological parents committed three or more crimes, so that the adoptees were most likely to become criminal, fully 75% of adoptees never committed a crime. This is very problematic if one is interested in preventing crime, because it implies that it may be difficult or impossible to predict with precision who is most likely to become criminal.

A large and fairly recent study obtained clear evidence of genes associated with criminal behavior.[109] Scientists examined the records of 862 men who were born out of wedlock in Stockholm between 1930 and 1949, and who were adopted by nonrelatives. A mountain of data was compiled, using records of the State Criminal Board, the Temperance Board, the National Health Care System, child welfare officers, and local hospitals all over Sweden, to obtain information on the criminality, alcohol abuse, medical problems, and social history of both adoptees and their parents. Adoptees were divided into four groups, depending on the criminal behavior of their biological and adoptive parents (Table 1). The largest group of adoptees came from a genetic background that included no criminality, and they were adopted into an environment of no criminality. However, about 14% of adoptees came from a background of high criminality, but were adopted into a law-abiding home, while another 8% came from a law-abiding background but were adopted into a family with high criminality. Finally, an unfortunate few came from a background of criminal behavior, and were adopted into an environment as bad. Nearly 40% of those adoptees with both a hereditary and an environmental exposure to criminal

Table 1
Criminal Behavior in Adopted Children[a]

Criminality of biological parents	Criminality of adoptive parents	
	Low	High
Low	3%	7%
High	12%	40%

[a]The frequency of criminal behavior in adoptive children, as a function of the criminal behavior of biological and adoptive parents. In this type of study, biological parents are the source of gene effects in adopted children, whereas adoptive parents are the source of environment effects on these children. This is the essence of a "cross-fostering study" design.[109]

behavior grew up to become criminals themselves. Both genes and the environment had an effect on criminal behavior of the adoptees, but the genes played a far more powerful role. Considering only those children placed into law-abiding homes, a child from a criminal background was more than four times as likely to become criminal himself as was a child from an honest background. But it was also found that little of the criminal behavior of adoptees could be explained solely on the basis of "criminal genes"; some interaction between genes and environment seemed to be required to induce criminal behavior. In fact, virtually all of the criminal behavior in Swedish adoptees was associated with alcohol use, so that a reduction in criminality could occur only if the rate of alcoholism was also reduced. On the basis of these findings, the heritability of criminal behavior in Sweden was estimated to be about 74%, which frankly seems likely to be an overestimate of the heritability of criminal behavior in the United States. Sweden is a very homogeneous culture compared to the United States; if environmental differences between children are minimized, as they are in Sweden, then individual differences must emerge from the genes. In other words, if everyone experiences the same environment and yet turns out differently anyway, then this implies that their

differences were inborn. In the United States, where there is often
a very great disparity in the environment experienced by children
of different social classes, it is anticipated that environment will
play a larger role in inducing person-to-person differences.

In general, alcoholic criminals tend to commit violent
crimes, while nonalcoholics tend to commit lesser property
crimes.[110] Alcoholic criminals tend to commit a large number of
violent offenses, whereas non-alcoholic criminals tend to commit
a small number of property offenses. The more a criminal drank,
the more violent crimes he committed, and criminal behavior
and alcoholism usually began within 2 years of each other.
Among mild alcohol abusers, about 22% of men were criminals,
while among severe abusers, nearly 67% were criminals. In fact,
the link between violent crime and alcohol is so strong that
scientists even proposed that it is not crime that is heritable, but
rather alcoholism, and that crime is simply a concomitant of
alcoholism. This seems like a restatement of the chicken-and-egg
controversy, though; property crime and alcoholism are both
heritable, and alcohol merely overwhelms some men with a
tendency to petty crime and makes them violent criminals
instead.

A hereditary tendency to crime is also strongly affected by
the sex of the potential criminal.[111] While the genes that predispose to crime appear to be comparable in the two sexes, the
congenital tendency to crime has to be more severe for a woman
to be equally affected. In fact, men seem to be more strongly
affected by genes, while women are more strongly affected by
the environment. The risk that a woman will become a criminal
is higher if she is reared in an urban environment or if she is
placed in an orphanage for a long time, whereas these factors are
both unimportant for men. A man is more likely to become
criminal if he has low social or socioeconomic status, or if he is
moved from foster home to foster home frequently as a child. It
is interesting that both placement in an orphanage and placement in multiple foster homes could impair the ability of the
child to bond with a parental figure. Alternatively, it could be
that children were deprived of an opportunity to bond with an

adult because they were somehow abnormal and prospective parents could sense this abnormality. In any case, the absence of a stable and loving relationship with an adult somehow predisposes the child to future criminal behavior.

Violence May Also Be Heritable

Despite strong evidence that violence is environmentally induced, there is also evidence that a tendency to violence is familial. According to the FBI, at least 18% of all homicides in 1990 involved family members killing one another.[102] The most common targets of murder in a family are fathers, sons, and especially brothers, and children under age 4 are more likely to be killed than are older children. Perhaps surprisingly, infants and small children are more likely to be killed by their mother than by their father. But, even though violence clusters strongly in certain families, this gives no clue as to how much of the clustering is related to shared genes and how much to shared environment.

The major risk factors for family violence are chronic alcohol abuse, social isolation of the family, and clinical depression of a family member, with some evidence as well of genes that directly transmit a proclivity to violence.[102] Yet, as we have seen in preceding chapters, chronic alcohol abuse, social isolation, and clinical depression all have a hereditary component, so it is quite likely that a tendency to violence would appear to be hereditary, even if there were no genes that cause violence *per se*. There are genes that indirectly predispose one to commit violence; a tendency to violence is associated with several personality disorders, and children who will become violent adults tend to have a specific personality profile.[112] Children with a restricted range of emotional responses, or who are unable to develop alternative approaches to interpersonal problems, or who consistently attribute hostile motives to others, are more likely to grow into violent adults. In addition, children prone to hyperactivity, impulsive behavior, and attentional deficits have a greater chance of growing into adults prone to violence. Such children tend to take chances, to be unable to delay gratification, to have little empathy, and to have a low IQ. These children often judge aggression to be an acceptable social response, and they tend to respond more aggressively than normal from a very early age. Their aggressive behavior is often judged by teachers, and even by other children, as inappropriate to the situation.

The risk factors that best predict which adolescents will later commit violent crime are no surprise[103]: drug use, poor school performance, low verbal IQ, childhood behavior problems, deviant peers, inconsistent parental supervision, lax discipline, a dysfunctional family, parental separation, childhood abuse, witnessing of violent acts, and poverty. But it is impossible to know which of these factors, and in which combinations, actually cause delinquency, and which are merely correlated with delinquency. This is a critical point, since ameliorating a causative factor would reduce delinquency, whereas ameliorating a correlative factor would likely have little or no effect.

Heritable Mental Illness May Predispose to Violence

Various types of mental illness have been linked to crime at one time or another, but the clearest link now is between attention deficit–hyperactivity disorder (ADHD) and criminality.[113] A group of 89 boys of various socioeconomic backgrounds, all of whom had ADHD, were recruited to a study before they had committed any crimes, then these boys were followed for 6 years. At the end of this time, boys with ADHD were compared to another group of boys of similar age and background who did not have ADHD. Boys with ADHD were more than twice as likely to be convicted of a crime, nine times more likely to be jailed for their offense, and ten times more likely to be jailed more than once. About 25% of all boys with ADHD were charged with a felony, whereas only 7% of the other boys were so charged. Yet virtually all this difference in arrest record could be accounted for by boys with ADHD who also had an antisocial personality disorder.

Since antisocial personality disorder is characterized by a wanton disregard for the rights and feelings of others, it is not surprising that this should increase the risk of arrest. Recent evidence suggests that antisocial personality disorder is somewhat more than half heritable, so this disorder is also likely to cluster in families. None of the boys with ADHD were jailed unless they also had antisocial personality disorder, whereas roughly a third of those children with both ADHD and antisocial personality disorder spent time incarcerated. Nearly two-thirds of the boys with both disorders eventually became known to the criminal justice system, either through an arrest, a conviction, or an incarceration. Antisocial personality disorder in children with ADHD also led to a multitude of problem behaviors, such as chronic truancy, verbal abuse of teachers, consistent disobedience, running away from home, destruction of property, and pathological lying.

Suicide is a violent behavior that clusters in certain families, and that is often associated with mental illness.[114] In fact, 90 to 95% of suicide victims suffer from some form of psychiatric illness at the time of death; approximately 45 to 70% are manic-depressive or suffer from clinical depression, while another 25% are alcohol-dependent. Scientists estimate that about 15% of all depressed patients die by suicide, and that depressed patients are 25-fold more likely than normal to take their own life. The fact that suicide is the eighth leading cause of death overall shows very clearly that suicide is a tremendous social problem in the United States.

A study of twins suggests that suicide is indeed familial, although the genes that predispose to suicide may be the same as the genes which predispose to mental illness.[115] A large data base, made up of white male twins born in the United States between 1917 and 1927, was searched to find out whether any of the twins had committed suicide. A total of 15,924 twin pairs were identified, and further analysis showed that there were 176 twin pairs in which at least one twin had died by suicide. Among these twins, it was found that more than 9% of identical twins were concordant for suicide, while fewer than 2% of fraternal twins were also concordant. Although the twin pairs had mostly been raised together, the fact that identical twins were more concordant than fraternal twins suggests that a tendency to suicide may be hereditary. Yet this effect is not a strong one; according to the traditional way of estimating heritability, the heritability of suicide is less than 15%. Of course, this low number is not trustworthy, because both twins and suicide are rare in the general population, and this double rarity makes it likely that the heritability estimate is in error. Furthermore, because mental illness is fairly common among suicide victims, it may be that concordance for suicide is simply the result of concordance for mental illness. Some scientists have also suggested that a person with poor impulse control might be predisposed to suicide. In this view, an inability to control impulsive behavior might conspire with stress, depression, or alcohol abuse, to increase the risk of suicide in some people. The most relevant genetic factor for suicide might, in fact, be an inability to control impulsive behavior.

Some scientists also maintain that brain structural changes are associated with a tendency to violence.[103] Positron emission tomography (PET) has been used to measure the metabolic rate of the brain, in murderers and in matched individuals who did not commit any violent crimes. PET scans of the murderers often show less metabolic activity than normal in the frontal part of the brain, which is believed to regulate aggressive impulses. Yet this reduction could be caused by any number of things, and there is no compelling reason to assume that this difference is important in determining a proclivity to violence. In fact, a reduction in brain activity in the frontal part of the brain could be caused by complications at birth, by abuse in childhood, or by head injury in adulthood.

Violence and Mutations of the Male Chromosome

One of the most firmly entrenched ideas in popular science is that men with an extra male sex chromosome have a strong tendency to violence.[116] The sex of a child is determined at conception by the father, who contributes either an X or a Y chromosome, whereas the mother can only contribute an X chromosome. If a child receives two X chromosomes, one from each parent, then the child will be female, and if a child receives an X and a Y chromosome, that child will be male. Under certain rare circumstances, a child may receive an X chromosome from the mother and two Y chromosomes from the father. It may be rather rare that such a child matures properly in the womb, but if birth does occur, then the child will have an extra male (Y) chromosome. Since many male sex characteristics are coded for by genes on the Y chromosome, it is not hard to see how the idea got started that an XYY male is a "supermale." Roughly 1 in 1000 newborn males have XYY chromosomes and so have the "super-male syndrome." Surveys of men in prison suggest that roughly 5 in 1000 prison inmates have this genetic makeup, so that supermales are apparently fivefold more common in prison than outside. Finally, a screening of men in prisons for the criminally insane suggests that the "supermale syndrome" is 19-fold more common among mentally ill men, or men who commit violent crimes. These findings are very controversial, at least in part because there is a tremendous range of variation in the frequency of the XYY mutation in different prisons. The frequency of XYY men varies so greatly from one prison to another that it often seems that scientists may simply be in error.

More recent evidence has shown that XYY men are less intelligent than normal, and that the violence associated with "supermales" is more a result of low intelligence than of an overabundance of any supposed male characteristics.[117] This conclusion emerged from a study that examined more than 31,000 men, born in Copenhagen between 1944 and 1947. From this large group of men, those men who were the tallest 15%

were identified, because past studies had shown that men with XYY chromosomes tend to be much taller than normal. By skewing the study to examine only the tallest men, scientists were trying to increase the odds that the study group would include men with XYY chromosomes. This left a total of nearly 5000 men to be studied. In admirably thorough and painstaking fashion, more than 90% of these men were interviewed and each contributed a blood sample which was tested for chromosomal abnormalities. From this monumental effort, a total of only 12 XYY men were identified, confirming that the abnormality is really quite rare. This alone should convince us that the "super-male syndrome" is not, in the broad scheme of things, critically important as an explanation of violence, since so few men have it. But, in comparing XYY men to normal XY men, it was found that XYYs were more than four times as likely to have a criminal record. Nearly 42% of the XYY men had some form of criminal activity in their background, although usually it was a property crime, rather than a violent crime. In addition, XYY men were substantially less intelligent than normal, and far more likely not to have completed their high school education. No evidence was found that XYY men were any more likely than normal to commit violent crimes, although several of the men had a long criminal record of petty property crimes. The most persuasive conclusion from this work is that XYY men have impaired intelligence, rather than enhanced aggression. According to the National Academy of Science, one of the most prestigious scientific bodies in the United States, the claim that XYY men tend to be involved in violent crime is still unproven, as late as 1993.

Very recently, a mutation was found in several violent male members of an unfortunate family in the Netherlands.[118] This mutation is particularly interesting because it maps to the X chromosome; because males have only one X chromosome, whereas females have two, a mutated gene on the X chromosome is far more likely to affect men than women. This is because women have an alternative, usually unmutated, copy of the gene on their second X chromosome, whereas men have no failsafe mechanism; if a gene is carried on the X chromosome, men will express it. Consequently, a mutation of a gene on the X chromosome is sex-linked, meaning that men express the trait, even though women may carry it. The X-linked mutation in question was first identified in a large Dutch family that had 14 male members afflicted with sex-linked mental retardation; scientists who were studying this family noticed that several of these men also displayed an unusual pattern of aggressive, aberrant behavior. Afflicted men had an average IQ of 85, well below normal but not severely retarded. The behavior of these men included odd, episodic fits of violent behavior, usually triggered by anger, but out of all proportion with the provocation. Aggressive behavior tended to cluster during periods of a few days, during which time these men would sleep poorly and have night terrors. Several of these men had impulsively attempted to murder a relative or acquaintance during a fit of anger, and several other men attempted arson. Inappropriate sexual behavior was also a problem; one man raped his sister, and several others were arrested for voyeurism, exhibitionism, or sexual harassment.

On closer study, it was found that these men were impaired in their ability to metabolize several different chemicals that transmit nerve impulses in the brain.[119] The metabolic impairment is a direct result of the X-linked mutation, since the mutation disables an enzyme called monoamine oxidase type A (MAOA). Because MAOA is deficient in these men, they are unable to properly metabolize three neurotransmitters (*i.e.*, norepinephrine, serotonin, and dopamine). As a result, these men excrete an abnormally high level of certain chemicals in their urine, which provides an easy way to make the diagnosis. Because these neurotransmitters had been implicated in impulsive violent behavior in the past, this study was very compelling; it tied theory and practice together, while providing an understanding of the strange behavior of these men.

It is a fascinating possibility that a single mutation could produce behavioral effects by causing an inability to restrain impulsively violent behavior. This idea fits in well with past work, and it is simple enough that it should be relatively easy to get at the truth. If a mutant MAOA gene can actually induce violence, then it may eventually become possible to ameliorate some violence by drug therapy or even gene therapy. But it is still too early to know whether this finding is true, because scientists have not yet had a chance to replicate or more fully explore the work. We do not even know the frequency of this mutation in the general population, so it is quite possible that some men who are neither retarded nor violent have the mutation. It is also very hard to explain why MAOA inhibitors, drugs that are widely used in the treatment of depression, have never been implicated in inducing aggressive behavior.

There are several separate lines of evidence that together imply that aggression and violence are intimately linked to individual heritable differences. Deficiency of brain serotonin (one of the three neurotransmitters that are reduced in men with MAOA deficiency) has been associated with both impulsive violent behavior and alcoholism in humans. There is up to a 50-fold variation between healthy individuals in the expression of MAOA, and this variation is familial. The enzyme MAOB varies to as great an extent as MAOA.[120] Low levels of MAOB can be found in people with alcoholism, manic-depressive illness, and certain personality disorders, as well as in suicide victims. Prozac, a drug that raises serotonin levels in the brain, has a calming effect and reduces impulsive behavior in people prone to violence. Convincing laboratory evidence shows that "knockout mice," which are unable to respond to serotonin, are very aggressive.[121] These mice do not show any obvious developmental or behavioral defects, yet if they are confronted with an intruder mouse, they attack that intruder more vigorously and much more viciously than normal. If it is true that human violence is associated with deficiencies in the ability to secrete or respond to serotonin, this may mean that violence is actually hardwired into some humans.

Can Biology Provide a Rationale for Human Violence?

Animal research, especially research with primates, provides a context for studies of human behavior. This does not mean that some humans are comparable to monkeys; in fact, all humans are somewhat similar to monkeys, and we would do well to learn what lessons we can from what sources we have. There can be no doubt that we as humans are not far removed from our primate cousins, so that insights from primates can be applicable to the human situation.

Research with monkeys has shown that serotonin levels in the brain are apparently modulated by the environment.[122] Among vervet monkeys, the highest-ranking male in the social hierarchy tends to have the highest levels of serotonin, while lower-ranking males have correspondingly lower levels of serotonin. But social level is not determined by serotonin level; in fact, it is the other way around. Serotonin levels in the blood seem to reflect subtle social cues, so that serotonin rises in a male monkey as that monkey ascends the social ladder. Conversely, if a dominant male suffers a precipitous loss of status, his serotonin level falls. The effect of serotonin is rather pacific for the dominant male, since he rarely is involved in physical violence unless his leadership role is directly challenged. It is monkeys with low serotonin levels that are most prone to impulsive violence, perhaps because they have the most to gain from violence. It has even been proposed that male monkeys of low social status respond to their lack of status by breaking social rules.

Can there be any doubt that human males compete for status, and that increased status brings with it access to all sorts of perquisites, including females? Men of high status are more likely to have women available to them, and power is an aphrodisiac, as Henry Kissinger observed. Men compete for social status using whatever means is available to them, whether that be closing a financial deal, completing a corporate takeover, or hitting someone over the head with a chair. A man will defend his honor in the best way he can, whether that means disemboweling an adversary in court or shooting someone. When men kill someone they know, there is usually an audience; in fact, violence is often a sort of performance.[122] Ritualized violence, in the form of sports or cultural activity, is popular in virtually every society. In our society, as in a primitive society, a man's reputation depends in part on a credible threat of violence.

Genes and Violence

In most cases, the human tendency to violence cannot be directly attributed to genes. With the exception of the X-linked MAOA mutation, it seems that violence itself is not heritable, although conditions associated with violence certainly can be inherited. Previous chapters have shown evidence that alcoholism, mental illness, and personality disorder are all heritable, to at least some extent, and all of these conditions are associated with an increased tendency to violence. Thus, while it may be too simplistic to think of "violent genes," there are some genes that are nonetheless associated with violence.

It is probable that all humans share genes that under certain circumstances lead to violence. In fact, evolutionarily speaking, it is virtually impossible to imagine that humans lack an innate capacity for violence; we are hunters as well as gatherers, and we have a very long history of internecine warfare. Competition for limiting resources is constant in a subsistence society, so it is likely that human groups have been fighting within and among themselves ever since such groups first formed. Feuding between different groups and societies probably began at the dawn of time and certainly continues to this day. Surely we as a species would have been extinguished long ago if we were unable or unwilling to resort to violence.

Therefore, the way to minimize violence in our society may be to accept that poverty, racism, social isolation, and economic inequity create circumstances in which violence becomes almost inevitable. By this logic, the best way to reduce inner-city violence would be to provide a nonviolent route to achieve social status. Educational access, availability of job training, attainable entry-level jobs with a reasonable prospect of advancement, access to adequate housing, freedom from pervasive and poisonous racism, and protection from crime would, without question, reduce the rate of inner-city violence. Biology provides a strong rationale for human violence, but it also suggests a course of action congruent with the best of human instincts.

14

Sex, Genes, and Testosterone

There can be no question that men and women are profoundly different; beyond the obvious physical differences, and the cultural differences, which are arguably added on late in the course of things like icing on a cake, there seem to be profound differences in the way the sexes relate to the world and to each other. It often seems that men and women are so different that we neither know the other's language; it is as if we were trying to function in a foreign land, while working with a bad translator. Even seemingly precise words can have different nuances and shades of meaning for the different sexes, so that what one thinks was said is not what was perceived.

These differences could obviously arise for many reasons. There is a profound difference in the way boys and girls are raised, and some would argue that the psychological and emotional distances between men and women arise from these cultural differences. This is a sort of radical feminist viewpoint; there are no inherent differences, just differences that are applied to the framework of a human being, like clay is applied to an armature. The other end of the spectrum is that there are profound and inborn differences between the sexes, and that the cultural differences between men and women arise from the genes, as surely as do hair color and height.

As usual, the truth is somewhere in between. There are inborn differences between men and women, just as there are inborn differences between all people. There is a sort of consensus male, an amalgam of those traits most common to men, just

239

as there is a consensus female. And these two mythical consensus beings differ. But the origin of their differences is a mystery that likely will remain so. Without doubt, the environment during upbringing accounts for some of the differences between the sexes, but equally without doubt, the genes play a role. The problem is that we will probably never fully understand what mix of genes and environment is responsible for the sex differences. Males and females are genetically different from the very instant of conception, and they are raised differently from the moment of birth. It will never be possible to use a split twin approach to study males and females, because fraternal twins who differ in gender always experience very different versions of the world. Similarly, studies of the effect of environment alone are impossible; some few children have been raised as members of the opposite sex, but there is always some deep-rooted pathology of the parents involved. Gene linkage studies are equally unlikely to provide insight, because there are very profound and very extensive differences between the genes of men and those of women. And finally, testosterone is the wild card; it is possible that women would be much more like men if they too were exposed to as powerful a force for anarchy and wildness. Consequently, real answers to the question of nature versus nurture can probably never be known, and this leaves us relatively free to speculate.

The Origin of Sex Differences

Males and females differ genetically from the very moment of conception. Each woman has two X chromosomes, and when that woman forms eggs, each egg gets an X chromosome. While these two chromosomes are never identical, they are very similar, in that each chromosome carries the same basic set of genes. Conversely, men have a Y chromosome as well as an X; this is what defines their maleness. The X and Y chromosome differ in the complement of genes they carry, with one of the major differences being that the Y chromosome carries a gene that induces males to develop testicles. When a male forms sperm, half the sperm get an X chromosome and half the sperm get a Y chromosome. Those sperm that bear an X chromosome combine with an egg, during the process of fertilization, to form a nascent girl child. And those sperm that carry a Y chromosome combine with an egg to form a nascent boy child. In other words, the particular combination of sperm with egg determines whether the newly fertilized egg is genetically male or genetically female at the instant of conception.

Thus, the genetic decision to be male or female is made on the basis of chromosomal baggage carried by the sperm. After the particular combination of egg and sperm has determined chromosomal sex, the process of determining maleness *vs.* femaleness begins. Chromosomal sex at conception determines gonadal sex, which is the sex that will become apparent as the sex organs form and mature. Gonadal sex is evoked by the action of a sex-determining gene on the Y chromosome; if the Y chromosome is absent, then a female is formed, whereas a functional Y chromosome will induce formation of testicles. Ultimately, gonadal sex determines the actual apparent sex of a person, because the sex organs secrete hormones during development, and these hormones have a profound effect on the growth of tissue. The two hormones most critical to the process of sex determination are testosterone, the male hormone, and anti-Mullerian hormone, a hormone that suppresses the growth of female sex organs. Both hormones are produced by the testes. The Y chromosome is thus a dominant determinant of sex because, when the Y chromosome is present, testes are induced and male sexual characteristics result. In the absence of the Y, testes fail to develop, and a female developmental pathway is followed. From a genetic standpoint, femaleness is a default setting, since no specific hormones are required for the formation of female genitals. Masculinization results from the ability of testosterone to induce male genitals and male secondary sex characteristics.

Yet this fairly simple system can go wrong, in ways that are revealing about the origin of sex differences. We have said that normally anyone with two X chromosomes is female, but every so often a man by physical appearance is found to have two X chromosomes.[123] Since, in the absence of a Y chromosome, this person should be female, something clearly has gone wrong. Such XX men have testicles and other external male genitalia, yet they are sterile and genetically female; this sort of sex reversal affects roughly 1 in 20,000 men. Apparently, this sex reversal occurs when a tiny piece of the Y chromosome carrying the sex-determining gene breaks off and reattaches itself to an X chromosome. Thus, even though two Xs are present, so is the male sex-determining gene, with the result that a male is formed. Yet the male formed is not entirely normal, perhaps because other essential male genes from the Y chromosome are missing.

There are other mutations that can produce genetic males unable to make testosterone, with the result that these "men" are pseudohermaphrodites.[124] This rare condition is one in which subjects have apparently normal female genitalia at birth, yet menstruation never occurs; instead, at puberty, the clitoris begins to grow into something resembling a micropenis. Careful examination of these subjects reveals that internally they are male; testes are present but undescended, and the pattern of internal plumbing is male in character. This strange combination of traits results because, although the subjects have an X and a Y chromosome and are therefore genetically male, the Y chromosome is abnormal. A gene involved in the synthesis of testosterone is carried on the Y chromosome, and mutation of this gene means that testosterone cannot be synthesized. Since pseudohermaphrodites cannot synthesize testosterone, they appear to be female because, in the functional absence of testosterone, the patients revert to femaleness as a default.

It might be that individuals with two X chromosomes and a Y are sort of an experiment of nature, since these individuals have all of the genes necessary to make both a normal male and a normal female. Conceptually, an XXY individual might be a female with added testosterone, or alternatively, the Y chromosome could be completely dominant over the X, so that the individual is a normal but perhaps sterile male. In fact, an XXY individual is psychosexually male and is said to have Klinefelter's syndrome; generally, these individuals have abnormally small testes that don't produce sperm, and they have reduced levels of testosterone, so that facial hair and other secondary sex characteristics are reduced in expression. Often these men are obese or have male breasts, and they can be mentally deficient and socially maladjusted. Men with Klinefelter's syndrome tend to be somewhat more prone to criminal behavior than normal, and are more likely to be of low intelligence.[117] In many ways, XXY men are like XYY men, although men with Klinefelter's are half as likely as XYY men to indulge in criminal behavior. Thus, various unusual mutations and syndromes that affect human sexuality all concur in showing that the Y chromosome is dominant in determining the development of sexual characteristics.

Males Are Genetically More Complex Than Females

In broad terms, men and women are identical from a genetic standpoint, with the exception of those genes located on the Y chromosome. This is because, although men have only one X chromosome, women express only one X chromosome. Early in the course of female development, one of the two X chromosomes is randomly inactivated, so that the genetic information carried on that chromosome cannot be used. The net result of X chromosome inactivation in women is that the individual cells of men and women both express only one X chromosome. However, males also express genes on the Y chromosome, so that men use a wider array of genes in day-to-day life.

In this sense, males are genetically more complex than females. Also there is genetically more that can go wrong with males; the mutation rate is six- to 11-fold higher in men than in women. The rate of human mutation was determined in a very clever way, by measuring the rate of production of new mutations among 45 families in Sweden.[125] All of these families have members afflicted with a blood disease known as hemophilia B (or factor IX deficiency), and all had donated blood in the past to help scientists learn more about this disorder. Hemophilia B is caused by mutations of a gene on the X chromosome, so it was possible to sequence the gene and characterize the mutation in each afflicted person and in his parents. Several of the mutations were found to be spontaneous (i.e., not present in the parents), and it was found that the rate of spontaneous mutation was 11-fold higher in men than in women. The rate of mutation may be faster in males than in females because more cell divisions are needed to produce sperm than eggs, or it may somehow be a result of exposure to testosterone. These results were confirmed by a more recent study that examined the relative rate of mutation in men and women, and concluded that males mutate about sixfold faster than females.[126]

The higher rate of mutation in men is a very mixed blessing, of course; it means that evolution is essentially male-driven, but it also means that males are more often the source of a harmful mutation. If one thinks of humanity as a culmination of billions of years of evolution, winnowed out by an equal period of harsh natural selection, then it becomes clear that humans are a rather finely honed machine. In the same sense that a blind man making random changes is unlikely to be able to make a functioning watch run more accurately, it is unlikely that a random gene mutation will improve on the function of the human machine. Consequently, most mutations are harmful, if not outright lethal. It is thus an onerous burden for males to be the source of most of the harmful mutations present in the world; the only possible comfort is that males are also, in an evolutionary sense, the wellspring of what makes us human.

Sexual reproduction is thought to have arisen because it allows organisms that reproduce in this way to mix their genes, and so to evolve more rapidly.[127] But the fact that males produce an order of magnitude more mutations than females creates a problem; it is unclear why females would risk combining their eggs with sperm that may well contain a mutation. In fact, computer simulation shows that the cost of male mutations to the female can actually exceed the benefits. If this is true, it would behoove females to reproduce without mixing of the genes. It thus becomes harder to understand how sexual reproduction became established in the human species in the first place. Many animals reproduce without sexual mixing of the genes, thereby conserving whatever particular genes they already have. While this would severely limit the rate of evolution, it would mean that a well-adapted animal would remain well-adapted, provided that the environment did not change. Yet this sort of computer analysis is somewhat specious; although it is possible to prove that females should reproduce without males, they can't, they don't, and all else is fanciful. Nevertheless, it is an intriguing problem to anyone interested in the costs and benefits of sex.

Physical Differences between Men and Women

There are some innate and fundamental differences between men and women that go far beyond the obvious secondary sex characteristics. In fact, there are differences between men and women in the form and content of particular structures in the brain. In truth, these differences are rather subtle, but they may be large enough to imply differences in function.

The brain volume of men is about 10% larger on average than that of women, but it has long been assumed that this difference lacks functional significance.[128] Since there are no measurable differences in general intelligence between men and women, the larger brain volume of men may simply be related to the fact that men tend to be larger than women overall. The ratio of brain size to body size in men and women is roughly equal, and the larger brain volume of men is at least partially offset by the fact that brain cells are more densely packed in certain parts of the brain in women. Careful measurement has shown that women have about 11% more neurons, per unit of brain tissue, in the posterior temporal cortex. Because this part of the brain is involved in language formation, it is possible that this physical difference could cause a functional difference in fluency. But the male posterior temporal cortex is still larger than the female, so that men have 13% more neurons overall in this part of the brain.

When one takes a closer look at certain small parts of the brain, other differences emerge which may have functional significance. For example, a small clump of cells in a part of the brain called the hypothalamus is substantially larger in men than in women.[129] Furthermore, there are more than twice as many nerve cells in this clump, or nucleus, in men. Although the function of this part of the brain is unknown, cells here may be involved in sexual behavior. This tentative conclusion is made by analogy, since it is well known that the hypothalamus is involved in sexual behavior in other mammals; activity of similar cells in monkeys is associated with sexual behavior or penile erection. The hypothalamus recently gained a great deal of notoriety because a very controversial report[89] (discussed fully in Chapter 11) claimed that the size of one nucleus in the hypothalamus is related to sexual orientation. The claim was made that gay men have a third interstitial nucleus that is intermediate in size between that of women and that of straight men, and that this is somehow causally related to their gayness. However, the report describing this alleged difference was so badly flawed that nothing is known for certain.

There is quite a bit of evidence that neural connections between the right and left halves of the brain are more extensive in women than in men. Functionally, the right and left halves of the brain are largely separate from each other, much as the New York City telephone exchange is separate from its counterpart in Los Angeles. The major interconnection between the left and right hemispheres is the corpus callosum, a thick bundle of nerves that bridges the fissure between the two hemispheres. The corpus callosum acts as an information conduit between the two hemispheres, much as a trunk line connects the telephone system of New York City to that of Los Angeles. For communication between the telephone exchanges of the two cities to be adequate, the trunk line connecting them must have a large capacity. By the same token, if the corpus callosum is too small, communication between the right and left hemispheres of the brain might be limited. Recent research has shown that women have a proportionally larger corpus callosum than do men.[130] A similar male/female difference is seen in the anterior commissure, a tiny tract of nerves that also connects the two hemispheres.[90] The commissure is 28% larger in cross-sectional area in women, even when corrected for differences in body weight. These structures are not related in any known way to sexual or reproductive behavior, so these physical differences are all the more intriguing. There appears to be less lateralization, or functional separation of the hemispheres, in women than in men, which may account for cognitive differences that will be discussed later.

The male/female differences in brain structure are compelling, because they imply that the apparent differences between men and women are based on real physical differences in the brain. If men and women had brains that were physically identical, it would be hard to argue that the differences between men and women derive from anything other than learning or conditioning. But the fact that there are observable physical differences in that most fundamental organ, the brain, implies that men and women really do differ from one another, in ways that cannot be attributed to the environment.

Physiological Differences between Men and Women

There also appear to be important differences between men and women in the metabolism of the brain. Blood flow to the brain can be measured while a subject is thinking or trying to solve a problem, and this can reveal something about brain physiology.[131] Total blood flow to the brain increases whenever the brain is at work; blood flow to the left hemisphere increases during a verbal task, while blood flow to the right hemisphere increases during a spatial task. Blood flow to the brain is higher, per unit volume of tissue, for women than for men, when performing either a verbal or a spatial task. The fact that the metabolic rate of brain tissue is higher in women is consistent with the observation that neurons are more densely packed in parts of the female brain, although the significance of this difference is not yet known.

The rate at which the brain uses glucose can also be measured, and recent technical advances make it possible to measure regional metabolic rate with a great deal of precision.[132] This was done by injecting willing subjects with a radioactive form of glucose, which is then taken up by brain cells as these cells do work. This was done in 37 men and 24 women as they lay, quiet but alert, within a positron emission tomography (PET) machine. No significant differences were found between men and women, when metabolic rate was averaged over the entire brain, but there were significant regional differences. Men had higher metabolism in those parts of the brain that are most involved in physical action, while women had higher metabolism in those parts of the brain used during symbolic activities such as speech. Metabolic activity was asymmetric for most regions of the brain, in both men and women, but the metabolic asymmetry was slightly greater in women. Because subjects in this study were not performing a mental task, it is not known whether there are sex-related differences in the metabolic rate of the brain tissue when it is active. In a broad sense, this study demonstrated that the similarities are much greater than the differences, when comparing men and women, but it is still the differences that are more intriguing.

Cognitive and Emotional Differences between Men and Women

It has been thought for some time that males and females differ in their cognitive skills, perhaps in a way that could give insight into the battle of the sexes. But a recent analysis of a huge mass of data demonstrates that, while male/female differences do exist, the differences are rather subtle.[133] During the period between 1960 and 1992, roughly a quarter-million children between the ages of 14 and 22 were tested for reading comprehension, vocabulary, and mathematical ability. A subset of these children was also tested for a range of more specialized skills and abilities, such as mechanical reasoning, perceptual speed, writing ability, spatial ability, knowledge of science and social studies, and general knowledge of machinery. All of these children were not tested in exactly the same way, but meta-analysis techniques were used to compare all children by the same criteria. The crucial difference between this study and all previous studies of general intelligence in children is that the recent study was a random sample of all children. This is a major difference because a test such as the Scholastic Aptitude Test (SAT) is usually administered to a very select group: those children finishing public school with good enough grades that they are contemplating a college career. Those children who never finish public school, or those children who finish but do badly, or those children who simply don't want to go to college, never have occasion to take the SAT, so SAT scores are not truly representative of the whole population. However, the quarter-million children analyzed in this recent study were randomly selected, to form a truly representative sampling of the entire United States.

Meta-analysis of this huge accumulation of data revealed a number of interesting differences between the sexes.[133] On average, females tended to perform somewhat better than males on tests of reading comprehension, perceptual speed, and vocabulary, whereas males performed somewhat better on tests of mathematics, social studies, science, and spatial ability. Females tended to score substantially higher than males on tests of writing ability, while males scored dramatically higher than females on tests of vocational aptitude (e.g., mechanical reasoning, knowledge of electronics, and automobile and shop information). The largest sex differences were found in subjects not traditionally taught in school; the male/female gap was enormous in knowledge of electronics or in automobile and shop information. Yet there was also a substantial difference between males and females in subjects that are presumably taught equally to both sexes; females scored nearly twice as well as males on tests of writing ability. However, the tests that showed the largest male/female differences were usually administered to relatively small subsets of children, so they are more likely to be in error. Overall, considering only those tests for which there is a great deal of information, the average difference between the sexes was small. But these differences also appear to be fairly stable over time; there was no sign of a narrowing gap in mathematical ability over the 32-year span of the study.

Because there was so much information on reading comprehension, vocabulary, mathematics, and perceptual speed of children, several very interesting points emerged from the data. In general, males show substantially more variability than females in the scores obtained on these tests. This may seem like a subtle (even boring) point, yet it may have major implications for our society nonetheless. To clarify the importance of variability in test scores, imagine that a large population of children is tested for mathematical ability. The test score obtained by each child can be plotted, to form the all-too-familiar bell curve. The most commonly obtained test score falls at the highest part of the curve, and in a very large population, this test score is usually very close to the average test score. The width of this curve is described by a number known as the variance. If the variance of a bell curve is very small, the bell curve actually looks like a spike, whereas if the variance is very large, the bell curve appears broad and squashed. Variance thus tells us about the distribution of test scores obtained by a population of children. In tests of mathematical ability, the variance of male test scores is substantially greater than the variance of female test scores. This means that male test scores fall across a broader range of values; while the average score obtained by males and females is very similar, there are more males who score either very well or very poorly.

The difference in variance between males and females has very important consequences if one is interested specifically in those children who score highest on a test. If one considers only the top 10% of scores obtained on a test of mathematical ability, there will be 30% more males than females in this group (i.e., the ratio of males to females will be 1.3). In the top 3% of mathematical test scores, males will outnumber females more than two-to-one (i.e., male to female ratio = 2.1). In the top 1% of mathematical test scores, the male-to-female ratio will be about 7. People who have careers in science and engineering are overwhelmingly more likely to have scored in the 90th percentile or better on mathematics tests in high school. Sex differences in variance imply that substantially fewer females than males will score in the top 10% of a test of mathematical or scientific ability. Thus, the achievement of equal representation of the sexes in techno-logical jobs of the future may be difficult, because there are one-half to one-seventh as many women as men who excel in the relevant abilities. In addition, job salary is known to be closely related to mathematical ability, so these dry-seeming numbers may actually predict future job earnings as well. In short, in a meritocracy based purely on mathematical ability, males would be expected to outnumber females nearly two to one in the most demanding jobs.[133]

Overall, men are consistently better at mathematical and spatial relationships than women, whereas women are generally better at verbal tasks. In most people, with the exception of those who are left-handed, the right side of the brain is specialized for processing of spatial information, while the left side is special-ized for processing of verbal information. Because men are more skilled at spatial or right-brain tasks, while women are more skilled at verbal or left-brain tasks, there has long been a suspicion that men and women use their brains in a fundamen-tally different way. But the latest evidence suggests that men and women may even differ in the way that they use their brain to accomplish the same task.

Scientists have tested men and women for differences in brain function, using an exciting new tool called functional magnetic resonance imaging (fMRI).[134] Conventional MRI yields a detailed anatomical image of the brain, but fMRI combines this with an indication of which part of the brain is active during the performance of a particular task. By carefully designing tasks to use different cognitive skills, scientists can determine which part of the brain is involved in processing visual information, or which part of the brain functions during a verbal task. This technique is new enough that it has not yet been widely used to investigate male/female differences, but this situation will probably change rapidly as more and more scientists begin to use fMRI. Scientists imaged 19 men and 19 women as each performed a set of four different tasks, which used four different sets of cognitive skills. All subjects were right-handed, so that any difference between right- and left-handers would not confound the results. Men and women were asked to do very simple tasks, such as to judge whether two words rhymed, or whether two words were of the same type. As subjects did these tasks, fMRI was used to show the location of the brain regions actually doing the work.

According to this study, men and women differ in the way that the brain is organized for language. During a verbal task, men tend to use a part of the brain on the left-hand side, at the front of the head, while women tend to use this region and a corresponding region on the opposite side of the head. In other words, men tend to do all the work of language on one side of their brain, while women use both hemispheres of their brain. These trends were fairly consistent among men and among women, so that the difference between the sexes was highly significant. This implies that the language skills of men and women differ in a physiological sense, and that this cannot be rationalized as a learned difference in language skills.

Another study, first reported in *The New York Times*, concluded that women are better than men at recognizing the emotional content of a facial expression.[135] Men and women were both presented with photographs of male and female models who were trying to portray a range of different emotions in their face. Each subject was asked to guess at what emotion was being portrayed, and then men and women were compared on the basis of their ability to correctly guess the modeled emotion. Men and women were both very good at determining the emotional content of a face; most of the time, a facial photograph was correctly classified as to the emotion displayed. Generally, both sexes were more sensitive to a happy face than a sad one, and both sexes were more sensitive to the emotion displayed in a male face than a female one. Yet, despite these broad similarities, there were some interesting differences as well. Overall, women were somewhat more sensitive to the facial display of emotion than were men, and women were especially sensitive to happy faces. Men were more sensitive to sad expressions in another man's face than were women, although they were less sensitive to sad expressions in a woman's face. Generally, a woman's face had to be quite sad for a man to notice, and subtle modulations of expression in a woman's face often went unnoticed. These findings are consistent with the general stereotype that women are more "emotionally responsive" than men, although the degree to which men were sensitive to other men is something of a surprise.

The Potent Role of Testosterone

Forty years ago, the leading theory of human sexuality postulated that humans are sexually neutral at birth, and that gender identity and eventual sexual role is established through differential rearing of boys and girls. This idea gained currency in the 1960s, because it implied that men and women were fundamentally the same, and that the manifest differences between them are merely cultural. But the theory of sexual neutrality has now fallen on hard times; a series of fascinating scientific studies of human sexuality have demonstrated unequivocally that human sexual identity is largely a function of hormonal exposure. A fetus exposed to the potent male hormone testosterone will, in the absence of some sort of physiological abnormality, become functionally male.

In fact, the effect of testosterone is so potent that it can literally overwhelm the environment to which a child is exposed.[136] An unusual gene mutation, which is moderately common in a remote region of the Dominican Republic, has shown that testosterone is able to ordain gender identity, despite cultural determinants to the contrary. This mutation causes genetic (i.e., XY) males to underproduce testosterone while in the uterus, so that their sex organs are female in appearance at birth. These boys were mistakenly identified as female and were raised as girls in a society with a fairly rigid segregation of sex roles. Yet these children underwent changes at puberty more typical of males: their voice deepened, they added muscle mass and body hair, and their sex organs became more typically male in appearance. Of 18 boys who were raised as girls but who later underwent male pubertal changes, 17 changed gender identity to be congruent with their hormonal status. Thus, an overwhelming majority of these boys raised as girls were able to defy their environment successfully, in order to match their gender to their genes. Admittedly, several of these patients had psychosexual problems after their gender transformation, but 16 of 18 were able to function as males without any medical intervention at all. This shows clearly that gender is not immutably set by the environment, and that testosterone exposure has far more effect than environment in determining the sex role at maturity.

The power of testosterone in determining "maleness" is shown even more clearly in the case of transsexuals, a kind of bizarre experiment in sexuality.[137] A transsexual is a person who is apparently normal, from an anatomic and genetic standpoint, but who feels that they are actually a member of the opposite sex. Transsexuals have an absolutely fixed idea that they are "trapped in the wrong body," and they may express a profound sense of loathing for their own physique. This condition should not be confused with homosexuality; some males who believe they are "female" are nevertheless attracted to women, while some women who switch their gender are nevertheless attracted to men. Transsexuality is more common than might be assumed;

it is estimated to affect roughly 1 in 20,000 men and 1 in 50,000 women; if these estimates are correct, there are more than 8000 transsexuals in the United States. Careful study of transsexuals has failed to identify any factors that could explain the condition: there are no known abnormalities of the genes, of the genitals, or of levels of hormone in the bloodstream. Many medical experts believe that transsexuality is a form of pseudohermaphroditism, manifested mentally rather than physically. Transsexuality is perhaps made somewhat less mysterious by experiments with animals, which show that inappropriate hormone exposure, especially during a critical window of time near birth, can induce inappropriate sexual behavior.

Many transsexuals wish to undergo sex reassignment surgery, in which the surgeon amputates the natural sex organs and reconstructs sex organs appropriate to the opposite sex. Needless to say, this surgery is highly controversial; some psychiatrists have argued that it is unacceptable to sexually mutilate a person, no matter how much they want it, given that their want arises from a psychiatric pathology.[138] The transsexual may feel that they are "a woman trapped in a man's body," but this may be similar to the anorectic who believes herself to be obese despite obvious emaciation. We do not perform liposuction on someone with anorexia nervosa, and there may be an equally good reason not to perform sex reassignment surgery on a transsexual. In fact, many psychiatrists now believe that transsexualism is an unfortunate term for the disorder, and would prefer that it be called gender identity disorder, to make it clear that the problem is mental rather than physical.

In any case, many transsexuals elect to undergo sex reassignment surgery. As a part of this process, patients are required to undergo treatment with high-dose hormones of the opposite sex, the rationale being that this process is more reversible than surgery, should the patient change their mind.[139] Since transsexuals are ostensibly healthy, prior to hormone exposure, this medical procedure provides a rare opportunity to gain insight directly into the effect of hormones on human behavior. A recent study followed 22 women who were about to try to become male; these women were given an extensive battery of psychological tests prior to hormone therapy, and then again 3 months after treatment with testosterone began. Hormone treatment of these women produced physical changes within only a few weeks; facial and body hair grew, the voice deepened, and the clitoris enlarged to a more phallic form. Surprisingly, the psychological changes were about as rapid, and the effect of testosterone on mental function was very revealing. The ability of these women to think in three dimensions improved strikingly, while verbal fluency dropped dramatically, without any measurable change in overall intelligence. In other words, testosterone increased their spatial perception, and decreased their verbal fluency, making these women cognitively more like men. About 77% of the female-to-male transsexuals improved in spatial ability, while 94% worsened in verbal ability. Thus, administration of testosterone to women caused a shift in their cognitive abilities to a pattern more typically male.

There were equally dramatic personality changes caused by testosterone treatment of female transsexuals.[140] Treatment of 35 female-to-male transsexuals with testosterone caused an increase in anger, irritability, proneness to violence, interest in sex, and ease of sexual arousal. At the same time, a group of 15 male-to-female transsexuals was treated with estrogen and with a chemical that blocks the effect of testosterone. As these genetic males received hormone treatments to make them more "female," they experienced a decrease in anger, aggression, and sexual arousability. Changes in mental function went hand-in-hand with the personality changes: female-to-male transsexuals became less verbal and more able to work with shapes, while male-to-female transsexuals became more verbal and less able to work with shapes. In addition, the male-to-female transsexuals reported more mood swings than did the female-to-male transsexuals. Finally, male-to-female transsexuals reported an almost complete cessation of interest in sex. In short, hormone treatment rapidly altered personality traits in a fashion that could have been predicted by anyone familiar with cultural stereotypes. Yet it is unlikely that these patients were simply "acting out" their newfound sex roles for two reasons: first, changes were rather small, not the exaggerated or caricatured changes that might be expected in someone play-acting; and second, psychological tests are constructed so that it is not always obvious what a sex-stereotyped response would be. These results show that humans are not cemented into their sex roles, but rather that continued hormone exposure continually reaffirms the sex role. Unfortunately, this also makes it seem that the tension between the sexes is not only inborn, but inescapable.

These results suggest that testosterone is a very powerful drug; in fact, continued exposure to testosterone is associated with aggression and may even lead to violence. This latter possibility is suggested by a study in which scientists measured the amount of testosterone in saliva from 89 male prison inmates, and found that the testosterone level measured in saliva was related to the type of crime committed.[141] Prisoners with low levels of testosterone tended to be in prison for crimes such as burglary or theft, while prisoners with higher hormone levels were more often imprisoned for violent crimes, including rape and murder. This is not to say that high levels of testosterone inevitably lead to incarceration; rather, testosterone is associated with aggression, and aggression is sometimes expressed in violent behavior. Yet aggression can also be channeled in more useful directions. It may well be that trial lawyers have higher levels of testosterone than normal, but trial lawyers have learned to channel their aggression in an acceptable manner. If men who become trial lawyers were not blessed with intelligence and educational opportunity, they might well find less-acceptable ways to express this aggression. Broadly speaking, it is perhaps more difficult to be male than female in our society, because there may be an inherent conflict brought on by living in a regimented society. Being male may involve a continual effort to resolve the tension between individual urge and societal imperative, between genes and environment. Evidence for this contention is seen in the higher incidence of virtually every behavioral pathology in men.

Genes and Sexuality

There is convincing evidence that males and females are fundamentally different, in terms of brain anatomy, brain metabolism, cognitive function, and even emotional response. Males appear to experience the world in a way profoundly different than females, and the different experiences appear to be rooted in measurable physiological differences. These differences are so pervasive that it is essentially impossible to imagine how they could result from environmental effects alone.

On balance, the greatest underlying difference between males and females is in the degree of exposure to testosterone. It is thus reasonable to assume that testosterone accounts for much of the actual difference between men and women. Yet testosterone is produced within the body because certain genes are expressed; men have high levels of testosterone in the bloodstream and show male sexual characteristics only because they have a Y chromosome.

Therefore, overt differences between the sexes are ultimately the result of gene differences. This is not meant to imply that differences between the sexes are due exclusively to the genes; it is simply that the role of the genes in producing sexuality is somewhat easier to understand than the role of the environment. There is, as yet, no reliable way to separate the influence of genes and environment on human sexuality; the workhorse of behavioral genetics, the split twin experiment, cannot be used to study sexuality. This is because identical twins are always the same sex whereas fraternal twins are often of different sexes. Furthermore, an individual is male or female from the very moment of conception, and males and females are treated differently virtually from birth.

Perhaps the clearest insight into the nature of sexuality has come from studies of transsexuals who are receiving hormone therapy, but these studies are still in their infancy. It will be years before scientists can replicate these studies and determine with confidence the relationship between hormone exposure and behavior. Although the absence of hard data leaves us relatively free to speculate, the act of speculation should make a scientist somewhat uncomfortable; the role of the scientist is to generate data that constrain the imagination, rather than speculation that inflames it.

15

Biology and Social Responsibility

Word of exciting new advances in molecular biology can be found almost every week in the newspaper, and many people have become accustomed to hearing that each latest advance will have revolutionary consequences. The Human Genome Initiative was sold to the American public so effectively that many people expect that knowing the chemical sequence of DNA will make us privy to the function of every gene, and so will enable us to cure every problem in the near future. There is a pervasive feeling that DNA decrees destiny, that the future of humanity is indelibly written in our genes. A new cynicism has emerged about psychiatry and clinical psychology; modification of human behavior is felt to be hopeless, since our behavior is a part of the fiber of our being. Nature seems to have overwhelmed nurture, and people are frankly skeptical that nurture has much impact on nature. *The Bell Curve* even contends that, since genes determine intelligence, and intelligence determines income, then class structure is somehow inherent in the genome. In this view, welfare programs cannot possibly succeed, because welfare recipients are somehow too flawed to benefit from this largesse in any substantive way. Predilection and predestination are confounded, and the social fatalism of the Middle Ages is resurgent.

Does DNA Decree Destiny?

If DNA really decreed destiny, then every cell of the human body should be alike; every cell arises from a single precursor

261

cell, the fertilized egg, and so each cell has an identical complement of genes. But each cell type in the body is distinctly different. Clearly, the separate developmental pathways followed by different cell types are profoundly affected by something other than simply the presence or absence of genes. In fact, genes are differently expressed in different cells; a brain cell is a brain cell because it expresses certain genes typical of a brain cell, and these genes are quite distinct from those expressed in a skin cell. As far as scientists know, a cell follows a particular developmental pathway because of chemical factors in its immediate environment. The constant chemical dialogue between cells during development thus induces some cells to follow a certain developmental pathway, whereas nearby cells may follow a completely different developmental pathway. Our understanding of this process is fairly rudimentary, but experiments suggest that each cell subtly alters the microenvironment of surrounding cells, so that adjacent cells may experience an environment that is quite unique. In a sense, the only reason that skin cells and brain cells are not identical is that they have each experienced a profoundly different environment during development.

Thus, differences in the environment during maturation can profoundly alter the fate of cells. This is shown clearly by spina bifida, a congenital malformation of the spinal column that affects a small fraction of newborn children. Because the bones of the spinal column are incompletely formed in spina bifida, the spinal cord itself may protrude from the body, and paralysis or death can result. Spina bifida is hereditary to some extent, but it is also strongly affected by maternal nutrition during pregnancy. A woman with a chronically poor diet, deficient in folic acid from fresh fruits and vegetables, is more likely to have a child afflicted with spina bifida. Conversely, a woman who uses vitamin supplements during pregnancy is much more likely to have a healthy child, even if spina bifida runs in her family. Apparently, cells that form the bones of the spinal column are nudged down one of two different developmental pathways, depending on whether or not there is enough folic acid available at the time that this developmental decision is made. Would it be too surprising, then, that a child can be nudged down several different maturational pathways by the environment? Perhaps a child with a potential for antisocial personality disorder can be nudged to normalcy by an abundance of love and acceptance. Alternatively, a harsh, unloving environment may increase the severity of a personality disorder, such that what might have been mildly problematic is made severely disabling.

Human behavioral development must be far more labile, far more prone to environmental effect, than physical development, simply because it is so much more unformed at the beginning. At birth, we all have recognizable rudiments of our final physical appearance, but no least shred of adult behavior. Since virtually all of our behavioral development occurs in an environment redolent with stimuli, we must assume that these stimuli have an effect. We have seen, time and again, that the best evidence suggests that human behavior is very roughly half the result of genes and half the result of environment. But many mysteries remain: What features of the environment are most important in the development of different behavioral traits? How is the expression of a trait impacted by the environment? Can a gene for a behavioral trait be switched on or off? To what extent can a behavioral trait be manipulated by control of the environment? And finally, what are the social consequences of our knowledge of behavioral genes?

What Are the Social and Economic Consequences of Human Behavior?

Various human behaviors, including violence, alcoholism, mental illness, low intelligence, and learning disability, create a crushing psychological and emotional burden for many members of our society. All too often, this burden falls most heavily on those segments of society that are already burdened by racial and financial discrimination. For example, between 1979 and 1981, there were 59,000 excess deaths among blacks in the United States, with an excess death defined as one that would not have occurred if the mortality rate for blacks was comparable to that for whites.[142] The major cause of these deaths was homicide; the age-specific homicide rate for young black males is five to ten times higher than for most other groups. But, no matter which group is most heavily affected, the burden is borne in part by everyone. This creates a societal imperative that cannot be ignored.

A National Research Council Panel on the "Understanding and Control of Violent Behavior" concluded that, as well as the obvious imperative to minimize the scars of violence for the individual, there is an enormous economic incentive for society to reduce violence.[102] Even when death or injury is avoided, the economic cost of violence to victims and to society is very high: the estimated economic loss resulting from an attempted or completed rape is $54,000, while a robbery costs society $19,200, and an assault costs $16,000. About 15% of these losses are financial, including the victim's monetary losses, society's cost for lost worker productivity, the cost of emergency response, and the administration of compensation. But 85% of these costs reflect a value imputed to nonmonetary losses, including pain and suffering, psychological trauma, and reduced quality of life. If the response of the legal system to violence is factored in, this adds an additional financial burden to society, in the form of costs for police, the criminal justice system, and private security agencies. Finally, there are also other consequences to society, such as the destruction of families and neighborhoods, the fortification of schools, homes, and businesses, and the deterioration and abandonment of community resources, such as parks and playgrounds. A 1% reduction in violent crime was projected to save Americans about $1.2 billion each year.[103]

To take another example of the societal cost of human behavior, learning disabilities cost the American taxpayer about $5.8 billion annually.[143] As of 1993, there were 2.3 million children in the United States diagnosed with a learning disability, and an additional 120,000 students are diagnosed every year. About 80% of these children have dyslexia, a heritable impairment of the ability to read. Learning disability was officially covered by the Individuals with Disabilities Education Act in 1968, and there are now more children with a learning disability than with speech impairment, mental retardation, or a physical handicap. Learning-disabled children cost about $8000 a year on average to educate, whereas other children cost about $5500, so this represents a substantial cost to the taxpayer. But there is no agreed-upon diagnostic test for this learning disability, the underlying causes are largely unknown, and there are few clues as to how to prevent or treat the problem.

Learning disability is generally diagnosed when a child's math or reading ability is substantially poorer than predicted on the basis of his IQ. Unless there is such a discrepancy a child might just be considered mildly retarded. Learning-disabled children typically have trouble matching a written letter to a spoken sound, so that they are less able to use a phonetic approach to reading. Recent results suggest that learning disability is associated with a variant form of information processing in the brain, and that this condition may afflict 20% of the school-age children in the United States. It is unknown to what extent the environment can affect the development of learning disability, but there are apparently more children with a potential for learning disability than are ever actually diagnosed. This may mean that certain features of the environment can moderate the severity of disability, and that some environmental features may even protect children from developing learning disability.

Should We Learn about the Heritability of Behavior?

The tremendous social and economic consequences of human behavior strongly compel us to learn about the heritability of behavior. Mental disorders are aggravated by the environment in ways we do not understand, but we do know that the economic burden they create is passed on to society. Similarly, drug and alcohol addiction, both of which are heritable, cost the United States untold billions of dollars in lost worker productivity, increased need for medical, legal, and welfare services, and the cost of interdiction. Learning disabilities are a growing problem for the educational system and a growing cost for the taxpayer. And finally, the average career criminal commits 10–20 major felonies in his lifetime, meaning that a relatively small number of individuals create an enormous burden for society as a whole.

Greater understanding of the heritability of human behavior would enable scientists to achieve several different but intimately connected goals:

1. Determine possible causes for behavioral problems. It will be impossible to develop effective treatment strategies for a behavioral pathology until the origin of the pathology is better understood.

2. Achieve greater success in treatment of problem behaviors. Merely acknowledging the role of genes in contributing to behavior would be helpful; many scientists have spent too long discounting their importance. It is unlikely that a behavioral problem can ever be ameliorated if treatment is guided by the mistaken belief that human behavior is exclusively the result of environmental causes.

3. Improve risk estimates for behavior problems. Parents who want children, but who want those children to be free of behavioral pathology, would be very grateful for a means to predict with accuracy the real risk of problems for their children. Many parents might choose not to have children if they knew that the risk of pathology was high, while other parents might be better prepared for the difficulties involved in raising children they do choose to have.

4. Facilitate early diagnosis of behavior problems. Early diagnosis of a child with a problem might facilitate an effort to prevent the development of full-blown behavioral pathology. Early diagnosis of a learning disability might mean that a child could receive remedial education sooner, so that the problem could be overcome before the child experiences repeated failure.

Should We Set Limits on What Information Is Obtained?

Human beings express between 50,000 and 70,000 genes at some point during their life. This means that the genetic information necessary to make a human, rather than a monkey or a mold, codes for only 50,000 to 70,000 different proteins. Already, more than 5000 of these genes have been analyzed to the point where the impact of a mutation is at least partially understood. Of this modest number of known genes, more than 100 are intimately related to mental function, such that a mutation of the gene leads to a lowering of IQ.[22] In many cases, this means that a person's intelligence has been reduced by a genetic condition that will eventually be treatable. Rather than complaining about the issues created by low intelligence, it would be vastly preferable to solve the underlying problem. This can be done only if scientists are free to investigate the human genome without restriction.

The Human Genome Initiative is a massive project, initiated within the last few years, which proposes to sequence every human gene. Scientists believe they will eventually be able to identify genes for manic-depression, schizophrenia, drug addiction, Alzheimer's disease, and various other human scourges. Many genes may contribute to the expression of each of these traits, so it will take an enormous effort to develop a better understanding of these genes before disease prevention or cure can be contemplated. But, while the real benefits of this project may not be realized for many decades, there can be no doubt that human health will eventually be improved by the project.

There may be some risk attached to gaining more complete knowledge of the human genome, yet it is also clear that ignorance is dangerous. This was demonstrated graphically in the case of a boy with undiagnosed Tourette's syndrome who was first noticed because of unusual sexual behavior. Tourette's syndrome is an odd familial illness, characterized by uncontrollable nervous tics. However, these tics are not simply muscular; the tics of Tourette's can involve uncontrollable verbal outbursts, involuntary echoing of words spoken by others, and very characteristic spasmodic outbursts of obscenity. The symptoms of this particular boy included gales of obscenity combined with grabbing his crotch. School authorities brought this child to the attention of the police, with the result that the father was accused of sexual molestation. Only after the child had been removed from his home was the proper psychiatric diagnosis made and the father cleared of charges. While this is an isolated example of a rare disease, and one that should nonetheless have been easily recognized by most medical authorities, the point remains: both ignorance and knowledge have dangers implicit.

Some have argued that knowledge of the origins of human behavior is somehow inherently racist. Yet the most virulent racism is practiced by the most ignorant people, so ignorance is no protection. To refrain from seeking knowledge simply because it may offend a minority group is a weak argument. If a member of a minority group has a genetic tendency, perhaps even one that places him in a still-smaller minority, it is ultimately in his own best interest to learn more. Yet, at the same time, it is dangerously naive to think that knowledge will always be used in an altruistic way; there can be no doubt that more complete knowledge of behavioral genes may create some new problems, or exacerbate some old ones.

What Are the Potential Consequences of New Information?

There are now genetic tests that can characterize the familial risk of breast cancer, colon cancer, and melanoma.[6] Other genetic tests, which can predict risk of Huntington's chorea, Alzheimer's disease, cystic fibrosis, or muscular dystrophy, have been available for several years. There is little doubt that genetic tests will eventually be able to predict the familial risk of behavioral pathology as well. In some cases, a genetic test does not provide good news, and therein lies a quandary. Serious consequences can result from bad news; several women who learned that they were at greatly elevated risk of breast cancer have attempted suicide. Roughly 10% of individuals who are told that they have either an increased or *decreased* risk for Huntington's disease subsequently suffer some serious psychiatric problems.[144] Most patients tested for Huntington's report that they have lower levels of distress and depression one year later, even if their worst fears were confirmed. But most patients also say that they needed counseling, no matter what the diagnosis. Thus, while it may be unethical to deny genetic testing to someone who wants it done, such a test can still exacerbate the problem.

Genetic testing can identify people who carry a gene for a trait that they do not themselves express.[145] Identifying such people can create new problems; though not themselves at risk, these people can still transmit the trait to their children. There are several key considerations that may help in determining whether or not to perform a genetic test:

1. How likely is it that test results are correct? If a genetic test has a high likelihood of being wrong, then clearly it is not useful. If some people with a genetic tendency to alcoholism are falsely told they need not worry, this is a major problem. But if someone is falsely told he is at risk, this is potentially devastating. Lives could be strongly impacted for no reason if a genetic test is frequently wrong. And there is no such thing as a perfect test; whenever a new genetic test is implemented, mistakes will be made.

2. Can anything be done if a person is at risk? If a genetic test reveals that someone will become senile in later years, yet nothing can be done to prevent this, then it may not be advisable to know. Unless treatment can actually prevent an outcome, then predicting that outcome may do far more harm than good.

3. Will action be undertaken because of genetic testing? If a couple wants to screen a fetus for a trait, yet they refuse to consider abortion as an option, how will testing help them?

4. Could knowing test results make things worse? If a genetic test makes it possible to discriminate against someone who has no symptoms, then testing is not desirable. Accurate testing will mean that those who are ill without symptoms can be identified, so that these people could be denied benefits or medical coverage. Results of a predictive test for Huntington's disease have already been used to deny medical insurance to dozens of people. If a test reveals that someone has a 50% risk of mental illness, half the people so identified will nevertheless remain healthy. Yet they may be denied medical coverage anyway, simply because they are at risk.

5. Why do we need to know test results at all? If a test could detect some, but not all, learning disabilities, then test results could be used to deny help to someone who really needs it. Should we help only those who can prove they need it? Is it humane to deny assistance to a person simply because he has a disability that cannot be detected by current genetic testing?

How Do Current Laws Relate to Genetic Liability?

Employers and insurance companies already discriminate against a person with a gene that confers an increased risk of disease. Employers prefer to hire employees who are less likely to have large medical bills, and insurance companies routinely deny coverage to those with a "preexisting condition." But a genetic tendency to schizophrenia could be considered a preexisting condition, in the same sense as elevated blood pressure is a preexisting condition predisposing one to develop heart disease.

In March of 1995, the Equal Employment Opportunities Commission (EEOC) determined that, under the Americans with Disabilities Act of 1990, genetic susceptibility to disease is a disability protected by law.[146] An illustrative example was given by the Commission: an applicant for a job has a familial susceptibility to colon cancer, but he is currently asymptomatic and may never develop the disease. A conditional offer of employment is made, but this offer is rescinded when the employer learns about the genetic predisposition to disease. This is, of course, a rational business decision, given that someone with a familial risk of colon cancer has a greater chance of increased health care costs, lower worker productivity, and poor work attendance. But, according to the EEOC, a company would be open to legal action for rescinding a job offer under such circumstances.

Genes that predispose one to a behavioral problem are conceptually similar to genes that predispose one to disease. Given that the EEOC considers disease genes to be protected by the Americans with Disabilities Act, it seems likely that behavioral genes are similarly protected. In a sense, this is similar to the protection given by laws against racial discrimination. These laws are simply intended to level the playing field, so that those born at a disadvantage do not remain so forever. Clearly, low IQ, learning disability, and alcoholism are genetic liabilities, and it may be that these conditions are already protected by law. But where do homosexual tendencies or a proneness to violence fit into this schema? What are the broad implications of this legal interpretation for those with various genetic liabilities?

What Can We Actually Do about Genetic Liabilities?

There are really only two possible positive responses to a gene that predisposes to some liability, whether that liability is a disease or an undesirable behavior; either that liability should be prevented or it should be treated. The only way to prevent a genetic liability is to prevent the birth of a child bearing that liability. Prevention of genetic liabilities will require several separate programs to be put in place:

1. Public education—so that people can learn more about which problems are hereditary and which genetic liabilities

they may be able to pass on to their children. This has already been done repeatedly and there have been extensive public education programs, e.g., about Tay-Sachs disease among Jews, about sickle-cell disease among blacks, and about cystic fibrosis among Europeans.

2. Genetic screening—to identify those parents most at risk of producing children with a genetic liability, or to identify those fetuses that already have a genetic liability. This is now done routinely for some medical problems, but it will eventually become possible to do this for a wide spectrum of behavioral problems as well.

3. Genetic counseling—so that those parents who may produce children with a genetic liability will be able to discuss their options. It is essential that genetic counselors be widely educated about genetic risks and options. Some parents who carry a gene for a particular behavioral trait may elect not to have children, for fear that their children will be affected, but these people need to know the full range of their options.

4. Viable and accessible birth control—to reliably prevent conception by parents who elect not to have children because of the risk of genetic liability.

5. Access to abortion—so that fetuses identified as bearing a severe genetic liability can be terminated. Clearly, abortion should not be thought equivalent to birth control, nor is it something to be taken lightly; it is not a desirable option at all. But it is a necessary option.

It is complete hypocrisy to simultaneously restrict abortion and oppose birth control, while also trying to limit access to medical care by the poor; parents identified as being at risk of producing children with a genetic liability may be left with no options at all. Too often, critics of abortion offer nothing at all to the severely handicapped or to unwanted children born when abortion is not widely available.

It will probably not be very long before the techniques of molecular genetics are so advanced that it becomes possible to show that virtually everyone has skeletons hidden in their genetic closet. The range of problems that have a possible genetic basis is so vast that almost certainly no one has a flawless genome. Preventing the birth of a child with a genetic liability may thus be impossible. In many cases, it may be more desirable to help an afflicted child to succeed in spite of their liability. Nonetheless, all children will not be able to overcome their liability; a child with severe mental retardation may be unable to succeed no matter how much training is given. But most behavioral traits or deficits are far more treatable than severe mental retardation. For these reasons, it may be more appropriate to treat than prevent genetic liabilities in many cases.

In order for a behavioral treatment program to succeed, those at risk must be identified when they are still young and amenable to therapy, and there must be an effective therapy available. This will only be possible if scientists can develop a means to predict which children are at highest risk, and if society deems effective therapy to be worth the cost incurred. But there are complex ethical issues involved in this; while children may be identified as being statistically more likely to commit a violent crime, it is clearly unethical to treat all of these children as potential criminals. Future generations of psychiatrists may classify men with a high probability of violence or a high rate of criminal recidivism as having a particular personality disorder, but the real challenge is to identify these men *before* they commit their crimes. If scientists can diagnose those most at risk of committing crime, and treat them before the damage is done, society may validate this as the best approach to reducing crime. But, if the "disease" is a behavior that is intolerable or damaging to society, will treatment become compulsory? This may mean, in essence, that some people will be convicted of a crime before they have committed it. Since treatment may well have undesirable consequences, this is something that cannot be taken lightly.

And unless treatment has a good chance of success, and an equally good chance of improving the lot of the treated person, then treatment is morally indefensible.

Can Human Behavior Really Be Modified?

Careful analysis of a wide range of treatments for various behavioral deficits or problems suggests that such methods can be strikingly successful.[147] This is something of a surprise; often when an approach to modifying human behavior is analyzed, there is a great deal of ambiguity. Some few studies are decidedly positive, others are merely suggestive, and some are convincingly null, but by far the majority of studies are hopelessly inconclusive. Recently, in an effort to shed light on this confusing situation, an enormous meta-analysis was undertaken.

Meta-analysis is a mathematical tool that allows the scientist to combine separate results from a range of different studies into a single large study. This enhances the strength of the data, in that the number of people studied is usually very much larger, so that small effects can be identified and so that more decisive answers can be given. A recent meta-analysis provided an overview of 302 earlier meta-analyses, so the new study was really a massive meta-meta-analysis. This enormous effort analyzed a range of treatments for human behavior, each of which intended to induce a change, whether emotional, attitudinal, cognitive, or behavioral. Included as a part of this analysis of treatments were such things as structured and remedial education, behavioral modification, psychotherapy, family and marital intervention, juvenile delinquency programs, smoking cessation programs, and treatment for the developmentally disabled. Each of these treatments is relatively mature, in the sense that it has attracted the interest of scientists for long enough that there are many separate research studies to summarize. Behavioral therapies such as primal scream therapy, crystal healing, or similar fringe therapies were not analyzed, simply because these therapies are too new or too unusual to have attracted much research attention from scientists.

Each separate behavioral intervention was analyzed using a statistic called the mean effect size. This is a way of standardizing the difference between a group that receives treatment and one that does not receive treatment, so that the effect of treatment can be measured. Mathematically, this is often somewhat difficult to calculate, because early studies may not publish all of the relevant details, but conceptually mean effect size is not difficult to understand. If mean effect size is positive, then treatment produces a beneficial outcome, whereas if mean effect size is negative, the treatment is actually harmful. A mean effect size close to zero indicates little or no difference from treatment, while a number close to or greater than one indicates a very large benefit from treatment.

As an example, suppose one wishes to determine whether or not the Head Start Program, a program of early childhood education for the poor and disadvantaged, has been successful. The most objective way to determine the success of Head Start is to measure the cognitive skill and ability of children who have gone through the program, and compare their test scores to the scores of similar children who have not gone through the program. The average score of the Head Start (or "experimental") children and of the other (or "control") children can be measured with accuracy using any of a number of different cognitive tests. If the average test score obtained by control children is subtracted from the average score obtained by Head Start children, this is a measure of the improvement produced by Head Start. To compensate for the fact that different scientists may use different cognitive tests, the average difference between experimental and control children can be standardized mathematically; this is simply a matter of dividing the net difference by a measure of the variability in the test score expected for any large population of children taking the same test (i.e., the standard deviation of the mean test score). As of 1993, 71 separate studies of the Head Start Program had been done, and the mean effect size for these 71 studies was +0.34. This means that, on balance, Head Start results in a substantial improvement in the test scores of children who complete the program.

Unfortunately, mean effect size does not have a simple intuitive meaning; it is incorrect to conclude that Head Start causes a 34% improvement in test scores (because the mean effect size was corrected for the standard deviation of the test used). But a mean effect size of +0.34 is substantial, given that a mean effect size as small as +0.10 is often statistically significant. As a point of comparison, the same approach can be used to calculate mean effect size for various medical treatments, all widely acknowledged to be effective. The mean effect size for bypass surgery is less than +0.2, that for chemotherapy of breast cancer is about +0.1, and that for streptokinase treatment of heart attack is less than +0.1. Thus, a mean effect size of +0.34 for Head Start amounts to a ringing endorsement of the program.

The massive scope of the new meta-analysis is shown by the fact that the Head Start Program was only one of the 302 separate interventions analyzed. Of these 302 interventions, only 7 were shown to be worse than no treatment at all. Therefore, we can conclude that virtually every type of behavioral intervention used today works to a greater or lesser extent. In fact, most behavioral interventions work very well; 90% of the meta-analyses of different interventions reported a mean effect size of +0.1, and 85% were +0.2 or larger. The mean effect size averaged over all of the separate studies was +0.5, meaning that in most cases those individuals who received intervention or treatment were substantially better off than those who did not. Overall, the mean effect size for behavioral change by psychological means was roughly comparable to the effect size for behavioral change by drug intervention. The only conclusion possible is that therapy to induce behavioral change is often surprisingly successful.

These results are very convincing because, in the final analysis, the conclusions are based on more than 9000 separate smaller studies, involving well over 1 million people. The mean effect size of +0.5 is roughly equivalent to comparing an untreated group, in which 38% of subjects report some degree of improvement, with a treated group, in which 62% of subjects report some improvement. This huge difference occurs in spite of the fact that interventions are often of short duration or occur late in a person's life. But several caveats are appropriate nonetheless. These results were averaged across many individuals, with many different degrees of problem severity and many different genetic tendencies or deficits. It is simply impossible, at this point, to predict whether a particular intervention will work for a particular person.

In addition, these results do not imply that every type of behavioral intervention works; this meta-meta-analysis only reviewed those psychological therapies that are mature enough that other meta-analyses had already been done. This means that radical new therapies, and therapies that are not widely practiced, could not be analyzed. Finally, it is possible that even crystal therapy is somewhat more successful than no therapy at all, because the placebo effect is generally so strong in psychological research. The placebo effect is basically the result of hope and positive expectation; if a person really believes that a particular therapy will help, that expectation often is a kind of self-fulfilling prophecy.

Despite all of these caveats and reservations, one can reach a strong conclusion from these data. Well-developed, mature interventions designed to modify personal behavior appear to be broadly effective, to a degree that would surprise many psychologists. Therefore, it is totally inappropriate to think of DNA as destiny. Even though many people may have a strong proclivity to harmful or undesirable behavior, these behaviors can almost certainly be ameliorated or corrected by therapy.

If the misery of our poor be caused not by the laws of nature, but by our institutions, great is our sin.
—Charles Darwin, *Voyage of the Beagle*

Epilogue

One cannot help but be somewhat frightened by a family cursed with schizophrenia or manic-depression or family violence. But the mind is very good at building barriers to keep frightening thoughts at bay. The barriers go up so quickly that one is usually neither frightened nor empathic for long. Similarly, it is hard to avoid feeling a little bit superior to a family suffering problems created by low intelligence or drug addiction or alcoholism. Seeing a family afflicted by any of these various genetic liabilities, one of the first barriers to go up is the conviction that this can't possibly happen to my family. We're too healthy or too lucky, too smart or too good, for something like this to happen. One can easily develop a certain cynicism; this terrible problem is affecting them, so they must in some way deserve it. Either they did something awful, or they foolishly exposed themselves to risk, or they are simply not as nice as they seem. But this kind of thinking is only a defense against the truly frightening realization that anything can happen to anyone, and that no one is truly safe.

Some families choose to bear children, despite having been afflicted generation after generation by a genetic liability, only to see their children suffer problems as well. One's first sympathetic response to a family like this may be attenuated by the feeling that these people should never have had children. Aren't they aware of the problem that has held them back all their life? Don't they care that their children may be similarly affected?

Don't they have any thought for the future? But this kind of thinking is also a defense against the unthinkable: any family can be similarly afflicted because we are all at risk.

Every child deliberately conceived was born because his parents were gambling that the child would be well and healthy. No parent can be blamed for gambling and losing, especially if they didn't really realize they were gambling. Perhaps the parents had a resolute faith that God would protect the child from harm. Likely the parents had no understanding of genetics, so they didn't realize that a child's options are limited by the parents' genes. Probably the parents didn't know that, for example, intelligence is 60% heritable, so a child is unlikely to be much smarter than his parents. Each parent gambles, whether they know it or not, that their child can overcome obstacles and limitations to grow up and live a good life.

In fact, we are all gambling, every day of our lives, and the fact that life is a gamble inevitably means that there will be some who lose, some who are born impaired. To maintain that a child somehow impaired should never have been born is essentially to deny reality. The choice was made, the child is here, and we cannot ignore them. We must help those with handicaps, because to do otherwise means that we are more handicapped than they. We as a nation seem to have lost our spirit of social responsibility, as we engage in divisive arguments about who is entitled to what. All perspective has been lost, in a mean-spirited diatribe against the poor, the unfortunate, and the disenfranchised. But no one born rich deserves their money, any more than the poor trapped in a ghetto deserve their poverty. It is simply an accident of birth. The role of blind luck and random chance is just too large for any of us to live our life without help.

People differ. Through no fault or inherent merit of their own, people are born stupid or smart, gay or straight, stable or schizoid. We can at times modify the conditions of our birth, but other times, try as we might, we are unable to change. Genes determine the range through which we can be modified, which is another way of saying, you can't teach a carrot to sing. People bear responsibility for the children they choose to bring into the world, but they bear no responsibility whatever for the conditions of their own birth. This is not to imply that a person bears no responsibility for what they become; we are all charged with doing the best we can with what we've got. But we are also charged with helping one another to achieve; the best we can become individually is intimately and irrevocably tied to the best that others can become.

References

1. Kimble, G. A., Evolution of the nature–nurture issue in the history of psychology, *Nature, Nurture, and Psychology*, ed. R. Plomin and G. E. McClearn (Washington, DC: American Psychological Association, 1993): 498.
2. Plomin, R., M. J. Owen, and P. McGuffin, The genetic basis of complex human behaviors, *Science* **264** (1994): 1733–1739.
3. Hall, J. C., The mating of a fly, *Science* **264** (1994): 1702–1714.
4. Wheeler, D. A., *et al.*, Molecular transfer of a species-specific behavior from *Drosophila simulans* to *Drosophila melanogaster*, *Science* **251** (1991): 1082–1085.
5. Gabriel, S. E., *et al.*, Cystic fibrosis heterozygote resistance to cholera toxin in the cystic fibrosis mouse model, *Science* **266** (1994): 107–109.
6. Steen, R. G., *Changing the Odds: Cancer Prevention Through Personal Choice and Public Policy* (New York: Facts on File, 1995).
7. Lewontin, R. C., S. Rose, and L. J. Kamin, *Not in Our Genes: Biology, Ideology, and Human Nature* (New York: Pantheon Books, 1984): 322.
8. Bouchard, T. J., *et al.*, Sources of human psychological differences: The Minnesota Study of Twins Reared Apart, *Science* **250** (1990): 223–228.
9. Martin, N. G., *et al.*, The power of the classical twin study, *Heredity* **40** (1978): 97–116.
10. Pickens, R. W., *et al.*, Heterogeneity in the inheritance of alcoholism: A study of male and female twins, *Arch. Gen. Psychiatry* **48** (1991): 19–28.
11. Egeland, J. A., *et al.*, Bipolar affective disorders linked to DNA markers on chromosome 11, *Nature* **325** (1987): 783–787.

12. Beckwith, J., A historical view of social responsibility in genetics, *BioScience* **43** (1993): 327–333.
13. Garver, K. L., and B. Garver, Eugenics: Past, present, and the future, *Am. J. Hum. Genet.* **49** (1991): 1109–1118.
14. Weindling, P., The survival of eugenics in 20th-century Germany, *Am. J. Hum. Genet.* **52** (1993): 643–649.
15. Shevell, M., Racial hygiene, active euthanasia, and Julius Hallervorden, *Neurology* **42** (1992): 2214–2219.
16. Provine, W. B., Geneticists and the biology of race crossing, *Science* **182** (1973): 790–796.
17. Horgan, J., Eugenics revisited, *Sci. Am.* **268** (1993): 122–131.
18. Marks, J., Historiography of eugenics, *Am. J. Hum. Genet.* **52** (1993): 650–652.
19. Needleman, H. L., *et al.*, Deficits in psychologic and classroom performance of children with elevated dentine lead levels, *N. Engl. J. Med.* **300** (1979): 689–695.
20. Baghurst, P. A., *et al.*, Environmental exposure to lead and children's intelligence at the age of seven years: The Port Pirie cohort study, *N. Engl. J. Med.* **327** (1992): 1279–1284.
21. Scarr, S., Developmental theories for the 1990s: Development and individual differences, *Child Dev.* **63** (1992): 1–19.
22. Plomin, R., The role of inheritance in behavior, *Science* **248** (1990): 183–188.
23. Lander, E. S., and N. J. Schork, Genetic dissection of complex traits, *Science* **265** (1994): 2037–2048.
24. Davies, J. L., *et al.*, A genome-wide search for human type 1 diabetes susceptibility genes, *Nature* **371** (1994): 130–136.
25. Bouchard, T. J., Genes, environment, and personality, *Science* **264** (1994): 1700–1701.
26. Mann, C. C., Can meta-analysis make policy? *Science* **266** (1994): 960–962.
27. Loehlin, J. C., Partitioning environmental and genetic contributions to behavioral development, *Am. Psychol.* **44** (1989): 1285–1292.
28. Sattler, J. M., Assessment of ethnic minority children, *Assessment of Children*, 3rd ed. (San Diego: Jerome M. Sattler, Publisher, Inc., 1992): 563–596.
29. Takahashi, J. S., L. H. Pinto, and M. H. Vitaterna, Forward and reverse genetic approaches to behavior in the mouse, *Science* **264** (1994): 1724–1733.

30. Steen, R. G., *A Conspiracy of Cells: The Basic Science of Cancer* (New York: Plenum Press, 1993): 427.
31. Marenberg, M. E., *et al.*, Genetic susceptibility to death from coronary heart disease in a study of twins, *N. Engl. J. Med.* **330** (1994): 1041–1046.
32. Sorensen, T. I. A., *et al.*, Genetic and environmental influences on premature death in adult adoptees, *N. Engl. J. Med.* **318** (1988): 727–732.
33. Jamison, K. R., Manic-depressive illness and creativity, *Sci. Am.* **272** (1995): 62–67.
34. Rao, V. S., *et al.*, Multiple etiologies for Alzheimer disease are revealed by segregation analysis, *Am. J. Hum. Genet.* **55** (1994): 991–1000.
35. Sellers, T. A., *et al.*, Lung cancer detection and prevention: Evidence for an interaction between smoking and genetic predisposition, *Cancer Res.* **52** (1992): 2694s–2697s.
36. Comings, D. E., *et al.*, The dopamine D2 receptor locus as a modifying gene in neuropsychiatric disorders, *J. Am. Med. Assoc.* **266** (1991): 1793–1800.
37. Herrnstein, R. J., and C. Murray, *The Bell Curve: Intelligence and Class Structure in American Life* (New York: The Free Press, 1994): 845.
38. Hunt, E., On the nature of intelligence, *Science* **219** (1983): 141–146.
39. Jensen, A. R., Spearman's g: Links between psychometrics and biology, *Brain Mechanisms: Papers in Memory of Robert Thompson*, ed. F.M. Crinella and J. Yu (New York: Ann. N.Y. Acad. Sci., 1993) 702: 103–129.
40. Arvey, R. D., *et al.*, Mainstream science on intelligence, *Wall Street Journal* Dec. 13 (1994).
41. Morin, R., The data that formed "The Bell Curve", *Washington Post National Weekly Edition* Jan. 16–22, 1995: 37.
42. Tambs, K., *et al.*, Genetic and environmental contributions to the covariance between occupational status, educational attainment, and IQ: A study of twins, *Behav. Genet.* **19** (1989): 202–222.
43. Plomin, R., and J. Niederhiser, Quantitative genetics, molecular genetics, and intelligence, *Intelligence* **15** (1991): 369–387.
44. Bouchard, T. J., and M. McGue, Familial studies of intelligence: A review, *Science* **212** (1981): 1055–1059.
45. Benbow, C. P., and J. C. Stanley, Sex differences in mathematical ability: Fact or artifact? *Science* **210** (1980): 1262–1264.

46. Fulker, D. W., J. C. DeFries, and R. Plomin, Genetic influence on general mental ability increases between infancy and middle childhood, *Nature* **336** (1988): 767–769.

47. Loehlin, J. C., J. M. Horn, and L. Willerman, Modeling IQ change: Evidence from the Texas Adoption Project, *Child Dev.* **60** (1989): 993–1004.

48. Bailey, J. M., and W. Revelle, Increased heritability for lower IQ levels? *Behav. Genet.* **21** (1991): 397–402.

49. Plomin, R., and J. C. Loehlin, Direct and indirect IQ heritability estimates: A puzzle, *Behav. Genet.* **19** (1989): 331–342.

50. Capron, C., and M. Duyme, Assessment of effects of socio-economic status on IQ in a full cross-fostering study, *Nature* **340** (1989): 552–554.

51. Pedersen, N. L., *et al.*, Separated fraternal twins: Resemblance for cognitive abilities, *Behav. Genet.* **15** (1985): 407–419.

52. Rousseau, F., *et al.*, A multicenter study on genotype–phenotype correlations in the fragile X syndrome, using direct diagnosis with probe StB12.3: The first 2,253 cases, *Am. J. Hum. Genet.* **55** (1994): 225–237.

53. Gershon, E. S., and R. O. Rieder, Major disorders of mind and brain, *Sci. Am.* **267** (1992): 127–133.

54. McGuffin, P., and R. Katz, The genetics of depression and manic-depressive disorder, *Br. J. Psychiatry* **155** (1989): 294–304.

55. Bertelsen, A., B. Harvald, and M. Hauge, A Danish twin study of manic-depressive disorders, *Br. J. Psychiatry* **130** (1977): 330–351.

56. Biederman, J., *et al.*, Evidence of familial association between attention deficit disorder and major affective disorders, *Arch. Gen. Psychiatry* **48** (1991): 633–642.

57. Kelsoe, J. R., *et al.*, Re-evaluation of the linkage relationship between chromosome 11p loci and the gene for bipolar affective disorder in the Old Order Amish, *Nature* **342** (1989): 238–243.

58. Detera-Wadleigh, S. D., *et al.*, Close linkage of c-Harvey-ras-1 and the insulin gene to affective disorder is ruled out in three North American pedigrees, *Nature* **325** (1987): 806–808.

59. Hodgkinson, S., *et al.*, Molecular genetic evidence for heterogeneity in manic depression, *Nature* **325** (1987): 805–806.

60. Baron, M., *et al.*, Genetic linkage between X-chromosome markers and bipolar affective illness, *Nature* **326** (1987): 289–292.

61. Berrettini, W. H., *et al.*, X-chromosome markers and manic-depressive illness, *Arch. Gen. Psychiatry* **47** (1990): 366–373.

62. Bebbington, P. E., *et al.*, The Camberwell Collaborative Depression Study. I. Depressed probands: Adversity and the form of depression, *Br. J. Psychiatry* **152** (1988): 754–765.
63. McGuffin, P., *et al.*, The Camberwell Collaborative Depression Study. II. Investigation of family members, *Br. J. Psychiatry* **152** (1988): 766–774.
64. McGuffin, P., and R. Katz, Genes, adversity, and depression, *Nature, Nurture, and Psychology*, ed. R. Plomin and G. E. McClearn (Washington, DC: American Psychological Association, 1993): 217–230.
65. Jones, P., *et al.*, Child developmental risk factors for adult schizophrenia in the British 1946 birth cohort, *Lancet* **344** (1994): 1398–1402.
66. Kendler, K. S., and C. D. Robinette, Schizophrenia in the National Academy of Sciences-National Research Council Twin Registry: A 16 year update, *Am. J. Psychiatry* **140** (1983): 1551–1563.
67. Sherrington, R., *et al.*, Localization of a susceptibility locus for schizophrenia on chromosome 5, *Nature* **336** (1988): 164–167.
68. Lander, E. S., Splitting schizophrenia, *Nature* **336** (1988): 105–106.
69. Kennedy, J. L., *et al.*, Evidence against linkage of schizophrenia to markers on chromosome 5 in a northern Swedish pedigree, *Nature* **336** (1988): 167–170.
70. Detera-Wadleigh, S. D., *et al.*, Exclusion of linkage to 5q11–13 in families with schizophrenia and other psychiatric disorders, *Nature* **340** (1989): 391–393.
71. Crowe, R. R., *et al.*, Lack of linkage to chromosome 5q11-q13 markers in six schizophrenic pedigrees, *Arch. Gen. Psychiatry* **48** (1991): 357–361.
72. Suddath, R. L., *et al.*, Anatomical abnormalities in the brains of monozygotic twins discordant for schizophrenia, *N. Engl. J. Med.* **322** (1990): 789–794.
73. Schellenberg, G. D., *et al.*, Genetic linkage evidence for a familial Alzheimer's disease locus on chromosome 14, *Science* **258** (1992): 668–671.
74. Corder, E.H., et al., Gene dose of apolipoprotein E type 4 allele and the risk of Alzheimer's disease in late-onset families, *Science* **261** (1993): 921–923.
75. Frith, U., Autism, *Sci. Am.* **268** (1993): 108–114.
76. Rutter, M., *et al.*, Autism: syndrome definition and possible genetic mechanisms, *Nature, Nurture, and Psychology*, ed. R. Plomin and

G. E. McClearn (Washington, DC: American Psychological Association, 1993): 269–284.

77. Gottesman, I. I., Origins of schizophrenia: Past as prologue, *Nature, Nurture, and Psychology*, ed. R. Plomin and G. E. McClearn (Washington, DC: American Psychological Association, 1993): 231–244.

78. Eaves, L., *et al.*, Genes, personality, and psychopathology: A latent class analysis of liability to symptoms of attention-deficit hyperactivity disorder in twins, *Nature, Nurture, and Psychology*, ed. R. Plomin and G. E. McClearn (Washington, DC: American Psychological Association, 1993): 285–311.

79. Brody, N., Intelligence and the behavioral genetics of personality, *Nature, Nurture, and Psychology*, ed. R. Plomin and G. E. McClearn (Washington, DC: American Psychological Association, 1993): 161–178.

80. Kagan, J., D. Arcus, and N. Snidman, The idea of temperament: Where do we go from here?, *Nature, Nurture, and Psychology*, ed. R. Plomin and G. E. McClearn (Washington, DC: American Psychological Association, 1993): 197–210.

81. Rowe, D. C., Genetic perspectives on personality, *Nature, Nurture, and Psychology*, ed. R. Plomin and G. E. McClearn (Washington, DC: American Psychological Association, 1993): 179–195.

82. Oldham, J. M., Personality disorders: Current perspectives, *JAMA Med. Assoc.* **272** (1994): 1770–1776.

83. Livesley, W. J., K. L. Jang, D. N. Jackson, and P. A. Vernon, Genetic and environmental contributions to dimensions of personality disorder, *Am. J. Psychiatry* **150** (1993): 1826–1831.

84. DeFries, J. C., D. W. Fulker, and M. C. LaBuda, Evidence of a genetic aetiology in reading disability of twins, *Nature* **329** (1987): 537–539.

85. Cardon, L. R., *et al.*, Quantitative trait locus for reading disability on chromosome 6, *Science* **266** (1994): 276–279.

86. Bailey, J. M., and R. C. Pillard, A genetic study of male sexual orientation, *Arch. Gen. Psychiatry* **48** (1991): 1089–1096.

87. Buhrich, N., J. M. Bailey, and N. G. Martin, Sexual orientation, sexual identity, and sex-dimorphic behaviors in male twins, *Behav. Genet.* **21** (1991): 75–96.

88. Gladue, B. A., R. Green, and R. E. Hellman, Neuroendocrine response to estrogen and sexual orientation, *Science* **225** (1984): 1496–1499.

89. LeVay, S., A difference in hypothalamic structure between heterosexual and homosexual men, *Science* **253** (1991): 1034–1037.

90. Allen, L. S., and R. A. Gorski, Sexual orientation and the size of the anterior commissure in the human brain, *Proc. Natl. Acad. Sci. USA* **89** (1992): 7199–7202.

91. Hamer, D. H., *et al.*, A linkage between DNA markers on the X chromosome and male sexual orientation, *Science* **261** (1993): 321–327.

92. Nowak, R., Fear and loathing in the U.S. military: Psychological explanations for homophobia, *J. NIH Res.* **5** (1993): 53–57.

93. Cloninger, C. R., Neurogenetic adaptive mechanisms in alcoholism, *Science* **236** (1987): 410–416.

94. McGinnis, J. M., and W. H. Foege, Actual causes of death in the United States, *JAMA* **270** (1993): 2207–2212.

95. McGue, M., From proteins to cognitions: The behavioral genetics of alcoholism, *Nature, Nurture, and Psychology*, ed. R. Plomin and G. E. McClearn (Washington, DC: American Psychological Association, 1993): 245–268.

96. Blum, K., *et al.*, Allelic association of human dopamine D2 receptor gene in alcoholism, *JAMA* **263** (1990): 2055–2060.

97. Gelernter, J., *et al.*, No association between an allele at the D2 dopamine receptor gene (DRD2) and alcoholism, *JAMA* **266** (1991): 1801–1807.

98. Gelernter, J., D. Goldman, and N. Risch, The A1 allele at the D21 dopamine receptor gene and alcoholism: A reappraisal, *JAMA* **269** (1993): 1673–1677.

99. Gejman, P. V., *et al.*, No structural mutation in the dopamine D2 receptor gene in alcoholism or schizophrenia, *JAMA* **271** (1994): 204–208.

100. Crabbe, J. C., J. K. Belknap, and K. J. Buck, Genetic animal models of alcohol and drug abuse, *Science* **264** (1994): 1715–1723.

101. Holden, C., A cautionary genetic tale: The sobering story of the D2, *Science* **264** (1994): 1696–1697.

102. Reiss, A. J., J. A. Roth, and Panel on the Understanding and Control of Violent Behavior, *Understanding and Preventing Violence* (Washington, DC: National Academy Press, 1993): 464.

103. Gibbs, W. W., Seeking the criminal element, *Sci. Am.* **March** (1995): 100–107.

104. Widom, C. S., The cycle of violence, *Science* **244** (1989): 160–166.

105. Kellerman, A. L., *et al.*, Gun ownership as a risk factor for homicide in the home, *N. Engl. J. Med.* **329** (1993): 1084–1091.

106. Sloan, J. H., et al., Handgun regulations, crime, assaults, and
     homicide: A tale of two cities, N. Engl. J. Med. 319 (1988):
     1256–1262.
107. Kellerman, A. L., et al., Suicide in the home in relation to gun
     ownership, N. Engl. J. Med. 327 (1992): 467–472.
108. Mednick, S. A., W. F. Gabrielli, and B. Hutchings, Genetic influences in criminal convictions: Evidence from an adoption cohort,
     Science 224 (1984): 891–894.
109. Cloninger, R. C., et al., Predisposition to petty criminality in
     Swedish adoptees. II. Cross-fostering analysis of gene–environment interaction, Arch. Gen. Psychiatry 39 (1982): 1242–1247.
110. Bohman, M., et al., Predisposition to criminality in Swedish
     adoptees. I. Genetic and environmental heterogeneity, Arch. Gen.
     Psychiatry 39 (1982): 1233–1241.
111. Sigvardsson, S., et al., Predisposition to petty criminality in
     Swedish adoptees. III. Sex differences and validation of the male
     typology, Arch. Gen. Psychiatry 39 (1982): 1248–1253.
112. Dodge, K. A., J. E. Bates, and G. S. Pettit, Mechanisms in the cycle
     of violence, Science 250 (1990): 1678–1683.
113. Mannuzza, S., et al., Hyperactive boys almost grown up. IV.
     Criminality and its relationship to psychiatric status, Arch. Gen.
     Psychiatry 46 (1989): 1073–1079.
114. Buzan, R. D., and M. P. Weissberg, Suicide: Risk factors and
     prevention in medical practice, Annu. Rev. Med. 43 (1992): 37–46.
115. Roy, A., et al., Suicide in twins, Arch. Gen. Psychiatry 48 (1991):
     29–32.
116. Hook, E. B., Behavioral implications of the human XYY genotype,
     Science 179 (1973): 139–150.
117. Witkin, H. A., et al., Criminality in XYY and XXY men, Science 193
     (1976): 547–555.
118. Brunner, H. G., et al., Abnormal behavior associated with a point
     mutation in the structural gene for monoamine oxidase A, Science
     262 (1993): 578–580.
119. Brunner, H. G., et al., X-linked borderline mental retardation with
     prominent behavioral disturbance: Phenotype, genetic localization, and evidence for disturbed monoamine metabolism, Am. J.
     Hum. Genet. 52 (1993): 1032–1039.
120. Craig, I., Misbehaving monoamine oxidase gene, Curr. Biol. 4
     (1994): 175–177.

121. Saudou, F., *et al.*, Enhanced aggressive behavior in mice lacking 5-HT(1β) receptor, *Science* **265** (1994): 1875–1878.
122. Wright, R., The biology of violence, *New Yorker* **March** (1995): 68–77.
123. Weil, D., *et al.*, Highly homologous loci on the X and Y chromosomes are hot-spots for ectopic recombinations leading to XX maleness, *Nature Genet.* **7** (1994): 414–419.
124. Geissler, W. M., *et al.*, Male pseudohermaphroditism caused by mutations of testicular 17β-hydroxylsteroid dehydrogenase 3, *Nature Genet.* **7** (1994): 34–39.
125. Montandon, A. J., *et al.*, Direct estimate of the haemophilia B (factor IX deficiency) mutation rate and of the ration of the sex-specific mutation rates in Sweden, *Hum. Genet.* **89** (1992): 319–322.
126. Shimmin, L. C., B. H.-J. Chang, and W.-H. Li, Male-driven evolution of DNA sequences, *Nature* **362** (1993): 745–747.
127. Redfield, R. J., Male mutation rates and the cost of sex for females, *Nature* **369** (1994): 145–147.
128. Witelson, S. F., I. I. Glezer, and D. L. Kigar, Women have greater density of neurons in posterior temporal cortex, *J. Neurosci.* **15** (1995): 3418–3428.
129. Swaab, D. F., and E. Fliers, A sexual dimorphic nucleus in the human brain, *Science* **228** (1985): 1112–1114.
130. Steinmetz, H., *et al.*, Sex but no hand difference in the isthmus of the corpus callosum, *Neurology* **42** (1992): 749–752.
131. Gur, R. C., *et al.*, Sex and handedness differences in cerebral blood flow during rest and cognitive activity, *Science* **217** (1982): 659–661.
132. Gur, R. C., *et al.*, Sex differences in regional cerebral glucose metabolism during a resting state, *Science* **267** (1995): 528–531.
133. Hedges, L. V., and A. Nowell, Sex differences in mental test scores, variability, and numbers of high-scoring individuals, *Science* **269** (1995): 41–45.
134. Shaywitz, B. A., *et al.*, Sex differences in the functional organization of the brain for language, *Nature* **373** (1995): 607–609.
135. Kolata, G., Man's world, woman's world? Brain studies point to differences, *New York Times* Feb. 28 (1995): C1–C7.
136. Imperato-McGinley, J., *et al.*, Androgens and the evolution of male-gender identity among male pseudohermaphrodites with 5α-reductase deficiency, *N. Engl. J. Med.* **300** (1979): 1233–1237.
137. Gooren, L., The endocrinology of transsexualism: A review and commentary, *Psychoneuroendocrinology* **15** (1990): 3–14.

138. McHugh, P. R., Witches, multiple personalities, and other psychiatric artifacts, *Nature Med.* **1** (1995): 110–114.
139. Van Goozen, S. H. M., *et al.*, Activating effects of androgens on cognitive performance: Causal evidence in a group of female-to-male transsexuals, *Neuropsychologia* **32** (1994): 1153–1157.
140. Van Goozen, S. H. M., *et al.*, Gender differences in behavior: Activating effects of cross-sex hormones, *Psychoneuroendocrinology* **20** (1995): 343–363.
141. Dabbs, J. M., *et al.*, Saliva testosterone and criminal violence in young adult prison inmates, *Psychosom. Med.* **49** (1987): 174–182.
142. Griffith, E. E. H., and C. C. Bell, Recent trends in suicide and homicide among blacks, *J. Am. Med. Assoc.* **262** (1989): 2265–2269.
143. Roush, W., Arguing over why Johnny can't read, *Science* **267** (1995): 1896–1898.
144. Wiggins, S., *et al.*, The psychological consequences of predictive testing for Huntington's disease, *N. Engl. J. Med.* **327** (1992): 1401–1405.
145. Motulsky, A., Predictive genetic testing, *Am. J. Hum. Genet.* **55** (1994): 603–605.
146. Culliton, B. J., Genes and discrimination, *Nature Med.* **1** (1995): 385.
147. Lipsey, M. W., and D. B. Wilson, The efficacy of psychological, educational, and behavioral treatment: Confirmation from meta-analysis, *Am. Psychol.* **48** (1993): 1181–1209.

Index